Water Environment Modeling

Water Environment Modeling

Clark C.K. Liu,
Pengzhi Lin, and Hong Xiao

CRC Press
Taylor & Francis Group
Boca Raton London New York

CRC Press is an imprint of the
Taylor & Francis Group, an **informa** business

First edition published 2022
by CRC Press
6000 Broken Sound Parkway NW, Suite 300, Boca Raton, FL 33487–2742

and by CRC Press
2 Park Square, Milton Park, Abingdon, Oxon, OX14 4RN

© 2022 Taylor & Francis Group, LLC

CRC Press is an imprint of Taylor & Francis Group, LLC

Library of Congress Cataloging-in-Publication Data
Names: Liu, Clark C. K., author. | Lin, Pengzhi, 1969- author. | Xiao, Hong
    (Professor and Vice Director of Hydroinformatics Institute of the State
    Key Laboratory of Hydraulics and Mountain River Engineering), author.
Title: Water environment modeling / Clark C.K. Liu, Pengzhi Lin, Hong Xiao.
Description: First edition. | Boca Raton : CRC Press, [2022] | Includes
    bibliographical references and index.
Identifiers: LCCN 2021021787 (print) | LCCN 2021021788 (ebook) |
    ISBN 9780367442446 (hbk) | ISBN 9780367442439 (pbk) | ISBN
    9781003008491 (ebk)
Subjects: LCSH: Environmental hydraulics—Mathematics. | Streamflow—
    Mathematical models. | Water quality—Measurement.
Classification: LCC TC163.5 .L545 2022 (print) | LCC TC163.5 (ebook) |
    DDC 627—dc23
LC record available at https://lccn.loc.gov/2021021787
LC ebook record available at https://lccn.loc.gov/2021021788

ISBN: 978-0-367-44244-6 (hbk)
ISBN: 978-0-367-44243-9 (pbk)
ISBN: 978-1-003-00849-1 (ebk)

DOI: 10.1201/9781003008491

Typeset in Sabon
by Apex CoVantage, LLC

Contents

Preface   xiii
Authors   xvii

1 Introduction   1

1.1   *Water resources and water quality management 1*
      1.1.1   *Hydrologic cycle 1*
      1.1.2   *Distribution of freshwater resources
              on the earth 3*
      1.1.3   *Hydrologic analysis for water quality
              management 4*
1.2   *Water environment modeling in water pollution control 7*
      1.2.1   *Evolution of water pollution control 7*
              1.2.1.1   *Minimum treatment requirements for
                        sewage discharge 9*
              1.2.1.2   *Permit system for urban sewage and
                        industrial wastewater discharge 9*
              1.2.1.3   *Federal government funding for the
                        construction of wastewater treatment
                        plants 10*
      1.2.2   *Water environment modeling fundamentals:
              (1) surface water 10*
      1.2.3   *Water environment modeling fundamentals:
              (2) groundwater 14*
1.3   *Water quality model applications 17*
      1.3.1   *Uncertainty in modeling analysis 17*
      1.3.2   *Basic procedures for water environment
              model application 19*
              1.3.2.1   *Mathematical formulation 20*
              1.3.2.2   *Mathematical solution 21*

        1.3.2.3   *Model calibration 21*

        1.3.2.4   *Model verification 21*

        1.3.2.5   *Model application 21*

1.4   *Prospects for water environment modeling 21*

    1.4.1   *Integration of traditional and linear systems modeling approaches 21*

    1.4.2   *Modeling with modern information technology 22*

    1.4.3   *Modeling with modern biological technology 22*

    1.4.4   *Modeling the fate and transport of emerging contaminates in water environment 22*

Exercises 24

References 24

2 Environmental hydraulics and modeling   **27**

2.1   *Mass conservation principle 27*

    2.1.1   *Mass continuity equation 27*

    2.1.2   *Mass transport equation 29*

2.2   *Molecular diffusion 30*

    2.2.1   *The basic molecular diffusion model and its analytical solution 31*

    2.2.2   *The random walk model of molecular diffusion 39*

    2.2.3   *One-dimensional advection–diffusion analysis 40*

2.3   *Turbulent diffusion and shear flow dispersion 42*

    2.3.1   *Turbulent diffusion and advection–turbulent diffusion equation 42*

    2.3.2   *Shear flow dispersion and advection–dispersion equation 49*

2.4   *Physically based water environment modeling 52*

    2.4.1   *Simulation of mass transport mechanisms 52*

    2.4.2   *Simulation of hydrodynamic mechanisms 53*

    2.4.3   *Mathematical formulation of physically based water environment models 54*

2.5   *System-based water environment modeling 55*

    2.5.1   *Mathematical formulation of system-based models 55*

    2.5.2   *Compatibility of physically based and system-based modeling approaches 56*

2.5.3   *System identification in linear system water environment modeling* 58
    2.5.3.1   *The method of physical parameterization* 59
    2.5.3.2   *The inverse method* 61
    2.5.3.3   *The method of system parametrization* 62

Exercises 62
References 64

3   Numerical methods for water environment modeling   67

3.1   *Analytical and numerical solutions of water environment models* 67
3.2   *Finite difference method* 68
    3.2.1   *Taylor expansion and scheme construction* 68
    3.2.2   *Truncation error and order of accuracy* 70
    3.2.3   *Consistency and convergence* 72
    3.2.4   *Stability* 72
        3.2.4.1   *Heuristic analysis of numerical stability* 72
        3.2.4.2   *Von Neumann analysis of numerical stability* 74
    3.2.5   *Schemes for convection equation* 75
        3.2.5.1   *Implicit methods (Laasonen method, Crank–Nicolson method, and ADI method)* 75
        3.2.5.2   *Explicit central-space methods (Lax method and Lax–Wendroff method)* 76
        3.2.5.3   *Upwind schemes* 77
        3.2.5.4   *Oscillation control methods* 78
        3.2.5.5   *Time-splitting methods* 78
    3.2.6   *Schemes for diffusion equation* 79
3.3   *Finite element method* 83
    3.3.1   *Background of finite element method* 83
    3.3.2   *Domain discretization* 83
    3.3.3   *Shape function* 84
    3.3.4   *Error minimization* 85
    3.3.5   *Matrix assembly* 86
    3.3.6   *Solution to linear system of equations* 87
        3.3.6.1   *Gauss–Seidel method and Jacobi method* 88
        3.3.6.2   *Successive over-relaxation method* 89
        3.3.6.3   *Conjugate gradient method* 89

3.4   *Numerical methods in QUAL-2K: A river and stream water quality model 90*

3.5   *Numerical methods in SUTRA: A groundwater flow model 93*

Exercises 96

References 98

4   **Ideal reactors and simple water environment modeling**   99

4.1   *Concepts and mathematical simulation of ideal reactors 99*

   4.1.1   *Hydraulic residence time distribution and mean hydraulic residence time 99*

   4.1.2   *Completely stirred tank reactor 100*

   4.1.3   *Ideal plug flow reactor 103*

4.2   *Laboratory determination of reaction kinetics and rate constants 105*

   4.2.1   *Zero-order reaction 106*

   4.2.2   *First-order reaction 107*

   4.2.3   *Second-order reaction 109*

   4.2.4   *Saturation-type reactions 110*

4.3   *Formulation and application of simple water environment models 113*

   4.3.1   *Detention ponds as CSTRs with time-variable inflow 113*

   4.3.2   *Lake water quality systems as CSTRs with steady-state flow 120*

   4.3.3   *River water quality systems as PFRs with steady-state flow 125*

   4.3.4   *Waste-assimilative capacity analysis of a water environment system as a CSTR or as a PFR 126*

Exercises 128

References 129

5   **Watershed hydrology and modeling for nonpoint source pollution control**   131

5.1   *Storm runoff and nonpoint source pollution 131*

   5.1.1   *Quantity and quality of storm runoff 131*

   5.1.2   *Watershed modeling for nonpoint source pollution control 133*

5.2   Linear systems approach to watershed rainfall–runoff
       analysis 135
       5.2.1   Unit hydrograph method 135
       5.2.2   Instantaneous unit hydrograph method 139
       5.2.3   Parameterization of instantaneous unit
                hydrograph 141
5.3   Geographic information system and watershed
       modeling 143
       5.3.1   Geographic information system 143
                5.3.1.1   Definition of GIS 143
                5.3.1.2   GIS applications 144
       5.3.2   The application of GIS in instantaneous unit
                hydrograph analysis 144
5.4   BASINS: A GIS-based water environment modeling
       platform 146
       5.4.1   The integration of BASINS and PLOAD model 146
       5.4.2   The integration of BASINS and HSPF model 150
Exercises 154
References 155

6   River water quality modeling   157

6.1   Effects of river hydraulic properties on its self-
       purification ability 157
       6.1.1   Spatial distribution of pollutants in a river 158
       6.1.2   Lateral turbulent diffusion 160
       6.1.3   Longitudinal dispersion 162
6.2   Effects of river reaction kinetics on its self-purification
       ability 164
       6.2.1   Depression of river dissolved oxygen content
                by waste deoxygenation 165
                6.2.1.1   First-stage carbonaceous wastes (CBOD)
                           deoxygenation reaction 168
                6.2.1.2   Nitrification and second-stage deoxygenation
                           reaction 171
       6.2.2   Replenishment of dissolved oxygen content by
                reaeration 174
       6.2.3   Dissolved oxygen variation in biologically
                active rivers 176
       6.2.4   River bacterial pollution and indicating
                microorganisms 178

6.3   Simplified river water quality models 178
       6.3.1   The Streeter–Phelps model 178
       6.3.2   The modified Streeter–Phelps models 185
                6.3.2.1   Enhancement of hydrodynamic
                           mechanism simulation – addition of
                           dispersion term 185
                6.3.2.2   Enhancement of reaction simulation 189
6.4   Comprehensive numerical river water quality
       models 189
       6.4.1   QUAL-2K model 189
       6.4.2   WASP model 193
       6.4.3   HEC-RAS model 193
6.5   Linear system approach to river water quality
       modeling 194
Exercises 199
References 203

7   Intensive river survey in river water quality modeling   205

7.1   Planning and execution of an intensive river water
       quality survey 205
       7.1.1   The procedure of conducting an intensive
                river water quality survey 206
                7.1.1.1   Preliminary survey plan 206
                7.1.1.2   Field reconnaissance survey 207
                7.1.1.3   River model formulation 207
                7.1.1.4   Final survey plan 207
                7.1.1.5   Execution of the final survey plan 208
                7.1.1.6   Survey report 208
       7.1.2   An example: intensive river water quality
                survey in Canandaiqua Outlet 208
7.2   River intensive water quality survey of hydraulic
       parameters 211
       7.2.1   River velocity and flow rate 211
       7.2.2   Time of travel 211
       7.2.3   Longitudinal dispersion coefficient 214
7.3   River intensive water quality survey of kinetic
       parameters 215
       7.3.1   Carbonaceous BOD deoxygenation
                coefficient 215
       7.3.2   Algal bio-productivity 217

7.3.3   Reaeration coefficient 220
    7.3.3.1   Measuring reaeration coefficient by tracer
         techniques 221
Exercises 225
References 228

8   Modeling of subsurface contaminant transport   231

8.1   Flow and contaminant transport in soils and
    groundwater 231
8.2   Mathematical simulation of flow and contaminant
    transport in soils and groundwater 231
    8.2.1   Transport simulation 231
    8.2.2   Water flow simulation 234
8.3   Modeling of contaminant transport in soils 235
    8.3.1   Numerical models 235
    8.3.2   Analytical models 236
    8.3.3   Integrative application of geographic information
         system and the simple index model of contaminant
         transport in soils 238
8.4   Application of linear systems theory in soil transport
    modeling 245
    8.4.1   Response function of a linear systems model 245
    8.4.2   Compatibility of traditional physically based
         soil transport model and linear systems soil
         transport model 246
    8.4.3   Conjunctive application of traditional
         transport model and linear systems model 251
8.5   Modeling of contaminant transport in groundwater 253
    8.5.1   Numerical models 253
    8.5.2   Analytical models 254
Exercises 257
References 260

9   Estuary, coastal, and marine water modeling   261

9.1   Linear systems modeling of estuaries 261
    9.1.1   Model formulation 261
    9.1.2   Model application 263
    9.1.3   Estimating the flushing time of a well-mixed
         tidal river 266

9.2   Coastal groundwater management and modeling 269
   9.2.1   Current issues in coastal groundwater
      management 269
      9.2.1.1   Land subsidence 269
      9.2.1.2   Seawater intrusion 270
      9.2.1.3   Sustainable yield 271
   9.2.2   Sharp interface approaches for seawater
      intrusion 272
   9.2.3   Variable density models for seawater
      intrusion 275
   9.2.4   Analytical models for sustainable yield 279
      9.2.4.1   Derivation of analytical coastal groundwater
         management model RAM2 279
      9.2.4.2   Determination of the sustainable yield of a
         coastal groundwater by RAM2 283
   9.2.5   Estimation of dispersion coefficient using
      data of deep observation wells 285
9.3   Marine water modeling 287
   9.3.1   Initial mixing of marine waste discharge 287
      9.3.1.1   Length-scale models 288
      9.3.1.2   Integral models 289
      9.3.1.3   Computational fluid dynamics models 289
   9.3.2   Initial mixing of artificially upwelled deep
      ocean water 290
Exercises 302
References 303

Index   309

Preface

Water environment modeling covers the formulation and application of mathematical models of water flow and mass transport. These mathematical models simulate the response of the water environment to waste inputs and are used as analytical tools for the planning and management of natural resources and environmental conservation.

The fate of waste substances in a natural or man-made water environment depends on its hydraulics and process kinetics. Early water quality models were formulated with simple hydraulics and relatively detailed process reaction. A noted example is the Streeter–Phelps model of dissolved oxygen (DO) variations in a river receiving oxygen-demanding organic wastes (Streeter and Phelps, 1925). The Streeter–Phelps model simulates the process reactions of organic deoxygenation and atmospheric reaeration by an ideal plug flow reactor (PFR). In the PFR, the transporting substances are instantaneously mixed in the lateral and vertical directions and the longitudinal mixing is due to advection only. Another noted example is the Vollenweider model of lake eutrophication (Vollenweider, 1976). The Vollenweider model simulates the lake hydraulics and relevant reaction processes of nutrient-stimulated algal growth by an ideal completely stirred tank reactor (CSTR). In the CSTR, the transporting substances are mixed with lake water immediately and completely.

As the severity of water environment problems intensifies and the scope expands, water environment models with simple hydraulics have become inadequate. Environmental hydraulics has evolved in response to the need for a more comprehensive analysis of water flow and substance transport in inland and marine waters. The study of substance transport in water environment by turbulent diffusion and shear flow dispersion started with the pioneering works of Taylor (1954). At present, most water environment models are conducted following Taylor's original concepts (Csanady, 1973; Fischer et al., 1979; Bear, 1979).

This book discusses the importance of both simple analytical models and comprehensive numerical models of various types of natural and man-made water bodies. The proper selection and application of these models are illustrated by examples and exercises, which are derived from actual case studies.

Simple and comprehensive models as introduced earlier are physically based models in the form of differential equations. In a physically based model, relevant hydraulics and reaction processes of a water environment system are represented explicitly by model parameters. The theory and application of linear systems models in the form of integral equations are also introduced in this book as an alternative modeling approach (Liu and Neill, 2002). In a linear system model, relevant hydraulics and reaction processes of the water environment system are represented implicitly by a system impulse response function.

This book is organized into nine chapters. Chapter 1 introduces environment models as analytical tools for water resources management and environmental conservation. Chapter 2 presents physically based water environment models based on environmental hydraulics and also presents alternative system–based water environment models based on the linear systems theory. Chapter 3 discusses numerical methods for solving governing partial differential equations (PDEs) of physically based water environment models. Chapter 4 discusses the analytical solutions and practical applications of simple water environment models as ideal reactors. Chapters 5, 6, 8, and 9 present the modeling of water environment systems including watersheds, rivers, subsurface water, and marine water, respectively. The success of physically based modeling lies upon the accurate estimation of the values of model parameters. To illustrate the importance of field data for accurate estimation of model parameters, the planning and execution of field intensive survey in a river to collect data for the estimation of model parameters are discussed in Chapter 7.

This book is prepared as a textbook for a senior or first-year graduate course on 'Water Environment Modeling' in civil engineering, environmental engineering, and other related academic fields. Chapters 1 through 7 can be used together as the textbook for a course on 'River Pollution Control and Modeling'. Chapters 1, 2, 3, 4, 8, and Section 9.2 in Chapter 9 can be used together as the textbook for a course on 'Groundwater Pollution Control and Modeling'. If this book is used for undergraduate teaching, Sections 2.3 and 2.5 in Chapter 2 may be excluded.

This book is also useful for practicing engineers and scientists as a reference book or as a supplement to the user manuals of models accessible by the public.

Although this book is jointly prepared by three authors, each chapter had a primary author or authors. Chapters 1 and 5 are by H. Xiao. Chapters 2, 4, and 7 are by C.C.K. Liu, Chapters 3, 6, and 8 are by P. Lin, and Chapter 9 is jointly written by the three authors.

It took more than 2 years to prepare this book from the lecture notes for a few undergraduate and postgraduate courses. These courses were taught at University of Hawaii, Sichuan University, National Taiwan University, Tongji University, and a number of other institutes.

The authors would like to acknowledge the kind efforts from the following people who have helped in various ways during the writing of this book. Special thanks go to Dr. Lian Tang, who has helped integrate different parts of writing into a clear draft and unify variables and equations. The postgraduate students Ms. Yaru Ren, Ms. Yuanyuan Tao, Mr. Runze Dong, Mr. Zhiming Ning, Mr. Huiran Liu, Ms. Yusi Wu, Mr. Wei He, Mr. Pengcheng Liu, Mr. Haiqi Fang, and Mr. Tao Liu have helped check equations and prepare figures and tables used in this book, and their efforts are gratefully acknowledged.

REFERENCES

Bear, J. (1979). *Hydraulics of Groundwater*. McGraw-Hill, Inc., New York.

Csanady, G.T. (1973). *Turbulent Diffusion in Environment*. REIDEL Communications, Springer, the Netherlands.

Fischer, H.B., List, E.J., Koh, R.C.Y., Imberger, J. and Brooks, N.H. (1979). *Mixing in Inland and Coastal Waters*. Academic Press, New York.

Liu, C.C.K. and Neill, J.J. (2002). Linear systems approach to river water quality analysis. Chapter 12 in *The ASCE Book: Environmental Fluid Mechanics – Theories and Application*. American Society of Civil Engineers, Reston, VA, pp. 421–457.

Streeter, H.W. and Phelps, E.B. (1925). *A Study of the Pollution and Natural Purification of the Ohio River, III. Factors Concerned in the Phenomena of Oxidation and Reaeration*. U.S. Public Health Service Publication, Washington, DC, No. 149.

Taylor, G.I. (1954). The dispersion of matter in turbulent flow through a pipe. *Proceedings of Royal Society London*, Series A. *Mathematical and Physical Sciences*, 223(1155), pp. 446–468.

Vollenweider, R.A. (1976). Advances in defining critical loading levels for phosphorus in lake eutrophication. *Memorie dell Instito Italiano di Idrobiologia*, 33, pp. 53–83.

Authors

Clark C.K. Liu, PhD, is Emeritus Professor and Former Chair of the Department of Civil and Environmental Engineering at the University of Hawaii. He earned his MS and PhD degrees from the University of Mississippi and Cornell University, respectively. He has served as a senior engineer and a senior scientist in the New York State Department of Environmental Conservation and as the Program Director of Environmental Engineering in the US National Science Foundation.

Pengzhi Lin, PhD, is Professor and Vice Director of State Key Laboratory of Hydraulics and Mountain River Engineering at Sichuan University, China. He earned his MS and PhD degrees from the University of Hawaii and Cornell University, respectively. He has worked at Hong Kong Polytechnic University and National University of Singapore as a postdoctoral fellow and an associate professor, respectively. He is the author of *Numerical Modeling of Water Waves* (CRC Press, 2008). He is Chief Editor of *Applied Ocean Research* and Associate Editor for *Journal of Hydro-Environment Research* and *Journal of Hydraulic Engineering*.

Hong Xiao, PhD, is Professor and Vice Director of Hydroinformatics Institute of the State Key Laboratory of Hydraulics and Mountain River Engineering at Sichuan University, China. He earned his MS and PhD degrees from Tianjin University and Florida State University, respectively. In 2017–2018, he was a visiting scholar at Kyoto University, Japan. His research interests include water-related disasters and environmental issues.

Chapter 1

Introduction

1.1 WATER RESOURCES AND WATER QUALITY MANAGEMENT

1.1.1 Hydrologic cycle

The hydrologic cycle, which describes the journey and balance of water on the earth's surface, below the ground, in the atmosphere, and back again (Figure 1.1), is an important concept in the discussion of generation, development, and protection of water resources.

Although the hydrologic cycle has neither a beginning nor an end, in order to discuss its components clearly, we usually start the cycle from evapotranspiration, which is the sum of evaporation from the land surface plus transpiration from plants. Water vapor that enters the atmosphere through the evapotranspiration processes condenses to form clouds. Under suitable conditions, tiny water droplets in the cloud further condense to form precipitation and return to the ground. It should be noted that precipitation includes rain, snow, hail, and so on. For convenience of discussion, it is collectively referred to as rain in this book.

The rain wets the soil and then begins to infiltrate into the underground. The infiltration rate varies with soil properties and land uses. For example, it can reach above 300 mm/h in forests, approximately 30 mm/h on farmlands, and even much lower in cities.

Overland runoff occurs when the rain rate exceeds the infiltration rate. As the infiltrated water enters the upper unsaturated soil, the water content increases. The water continues to flow slowly downward into the saturated layer, becoming groundwater and flowing slowly to the outflow area. The groundwater outflow area includes mountainous springs, riverbeds and lake bottoms, and seeps near the sea.

After the water enters rivers and lakes through the surface runoff and groundwater flow process, it continues to flow into the ocean. Evapotranspiration brings water back into the atmosphere from rivers, lakes, and oceans, completing the hydrologic cycle.

DOI: 10.1201/9781003008491-1

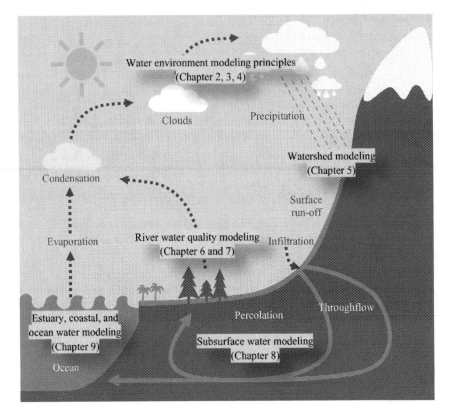

Figure 1.1 Hydrologic cycle and the scope of water environment modeling.

The two components, evapotranspiration and precipitation, in the hydrologic cycle can be viewed as a huge desalination project. The energy required to drive this project comes from solar radiation. The total amount of solar radiation energy received by the Earth is about 2.0 million TW (10^{12} W), of which 1.2 million TW reaches the ground, as presented in Table 1.1.

One-third of the solar radiation energy that reaches the ground (400,000 TW) is consumed by the evapotranspiration process of the hydrologic cycle (Overman, 1969), while the total energy consumption by the world population is about 18 TW (International Energy Agency, 2015). Table 1.1 also shows that the global energy loss from rivers to oceans is about 80 TW, among which less than 1 TW has been converted into hydropower. Building dams to develop hydropower can alter the ecological environment; therefore, further massive hydropower development is limited. Energy–water nexus is one of the biggest problems faced by mankind in the 21st century.

Table 1.1 Solar radiation energy that drives the hydrologic cycle

Energy Distribution	Value (TW)
Solar radiation energy received by the earth	2.0 million
Solar radiation energy reaching the ground	1.2 million
Energy consumed by evapotranspiration	400,000
Energy lost in the river	80
Energy converted into hydropower	1

As analytical tools in managing water resources and conserving ecological environment, water environment models are of great importance in meeting the challenge of energy–water nexus.

When water evaporates from the earth's surface, it absorbs heat and reduces ambient temperature, whereas it releases heat and increases temperature when it condenses. Therefore, the hydrologic cycle is closely related to the climate. Various economic activities of human beings in the modern era have greatly changed the solar radiation energy absorbed by the earth's surface, resulting in the problem of global warming. Global warming not only increases the average temperature on the earth but also changes the temporal and spatial distribution of global precipitation, which has a great impact on the development and management of water resources.

1.1.2 Distribution of freshwater resources on the earth

A large amount of water flows in different components of the hydrologic cycle, and its total volume is difficult to calculate precisely. Table 1.2 lists the estimated values that are commonly cited. Tm^3, a huge volume unit, is used in the table. One Tm^3 is equal to one million Mm^3, one Mm^3 is equal to one million cubic meters, and one cubic meter is equal to 1,000 liters. Table 1.2 shows that the total annual rainfall on the land is 120 Tm^3, and the evaporation amount is 80 Tm^3. The difference between the two is 40 Tm^3, which is the amount of fresh water flowing from the land to the sea every year and is also the amount of fresh water available to human beings theoretically. This huge supply of fresh water far exceeds human needs. However, there are still many regions in the world facing serious water shortage. Although the total amount of fresh water is not an issue, the lack of water resources is due to the uneven distribution of fresh water in space and time so that they cannot be effectively developed and used. In addition, human activities cause environmental pollution and destroy the originally good water sources.

To manage water resources properly, it is necessary to have a deep understanding of the flow speed in each part of the hydrologic cycle. Table 1.3

Table 1.2 Average annual water volume in the hydrologic cycle

Evaporation/Evapotranspiration		Rainfall	
From the ocean surface	420 Tm³	On the ocean surface	380 Tm³
From the ground	80 Tm³	On the ground	120 Tm³
Total	500 Tm³	Total	500 Tm³

Table 1.3 Percentage distribution of water and flow speed in the hydrologic cycle

Hydrologic Cycle Process	Earth Water Distribution (%)	Flow Speed (Order of Magnitude)
Atmosphere	0.001	100 Km/d
Surface water	0.02	10 km/d
Groundwater	0.52	1 m/d
Glacier	1.88	1 m/d
Ocean	97.58	....

shows the water distribution and typical flow speed in each process of the hydrologic cycle. It can be seen from Table 1.3 that the flow speed in the groundwater layer is only about 1/10,000 of that in the surface water layer (river). Because the speed of groundwater flow is extremely slow, the groundwater is basically a natural reservoir, which can be used to compensate the seasonal variation of rainfall, and becomes a good source of water. However, once the groundwater is contaminated, the pollutants are difficult to be removed due to the slow motion.

1.1.3   Hydrologic analysis for water quality management

The water in the rivers comes from two sources, namely, surface runoff in the watershed and baseflow from the groundwater. In a heavy rainfall event, the surface runoff generated in a watershed can lead to the flooding of the downstream reaches. Therefore, an important topic in the early application of hydrology is flood management. The main purpose is to estimate the runoff generated by a watershed when heavy rainfall occurs. The estimated runoff is also used as the design discharge for various types of flood prevention structures. To estimate the runoff, the Rational Formula is usually used for the urban drainage design, while the Unit Hydrograph is usually adopted for the design of large water conservancy facilities such as dams.

The analysis of flood hydrograph provides the design discharge for the plan and design of flood control facilities. It is an important topic in engineering hydrology though it was originally not related directly to water pollution prevention. In recent years, nonpoint source pollution has become the focus of water pollution prevention and control, which makes water quality models of both water quantity and quality capabilities become a powerful analytical tool in practices of water pollution prevention. Common water quality models for watersheds such as Better Assessment Science Integrating Point and Nonpoint Sources (BASINS)/Hydrologic Simulation Program Fortran (HSPF) also adopt the unit hydrograph method. The topic of water quality analysis for watersheds will be discussed in detail in Chapter 5.

In a dry season, if it does not rain for a long time in a watershed, there will be no surface runoff into the river. In such case, the river flow comes solely from the groundwater. This type of flow is called the baseflow, which is typically small and stable. The self-purification capacity of a river is directly related to its discharge. In the dry season, as the baseflow is small, the self-purification capacity is low, and consequently, the pollution is most likely to occur. Therefore, in the early practice of water pollution prevention and control, especially in calculating the degree of treatment for urban sewage and industrial wastewater, it was assumed that the receiving river is in a dry season state. In applied hydrology, the flow characteristics of a river in a dry season can be analyzed using the flow duration curve method and the low flow frequency curve method.

When drawing the flow duration curve for a river, it is necessary to replace the discharge record in descending order, divide it into several zones, and calculate the number of days in each zone. The resulting flow duration curve shows the percentage of time when the river discharge is equal to or below a certain value (Figure 1.2). Figure 1.2 compares the flow duration curves between two rivers. The catchment areas of the two rivers are close, but the characteristic of discharge in dry season is quite different. As shown in Figure 1.2, River 1 at point A can provide the minimum of 1.0 m³/s of water supply in the entire year, while the percentage of days to provide the same water supply for River 2 is only 50% (point B), although River 2 does provide more water in a rainy reason. Therefore, if a city needs a stable water supply of 1.0 m³/s, River 1 offers a better choice than River 2.

If the discharge record for a river over multiple years is available, the flood frequency curve of the river can be calculated by taking the annual maximum discharge and conducting frequency analysis. Similarly, if the annual minimum discharge is taken, the low flow frequency curve of the river can also be obtained through the frequency analysis. The main difference between the two is the duration of extreme values (maximum or minimum). Generally, when calculating the maximum discharge on the flood frequency curve, only its value is of concern, whereas when calculating the

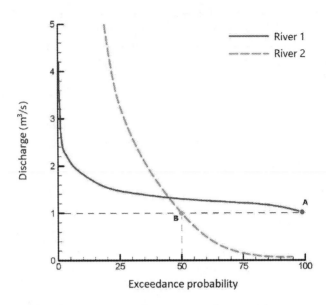

Figure 1.2 River flow duration curve and its water supply potential.

minimum discharge on the low flow frequency curve, both its value and duration should be considered. This is because for a flood event, such as dam or levee breach, even if the time duration is very short, it will cause serious damage. On contrast, for water supply and water pollution control, if the duration of low flow is short, it will not cause a big problem. Therefore, the flood frequency curve of a river is a single curve, while the low flow frequency curve is a series of curves, and each curve corresponds to the recurrence of low flow with a certain duration (e.g., 30 d). Figure 1.3 shows the low flow frequency curves of a typical stream, with the low flow duration ranging from 7 to 120 d. It can be seen in Figure 1.3 that for the recurrence interval of 10 years, the average low flow discharge with the duration of 7 d is about 0.015 m^3/s, while with the duration of 30 d, it is about 0.05 m^3/s.

The Federal Environmental Protection Agency of the United States (USEPA) stipulates that the minimum average discharge for a 7-d duration with a reoccurrence interval of 10 years (commonly known as MA7CD10) must be utilized, when applying a water quality model to determine the required degree of sewage treatment. In other words, when using a river water quality model to calculate the amount of sewage that can be discharged, the design flow of the receiving river should be set at a low flow state of MA7CD10.

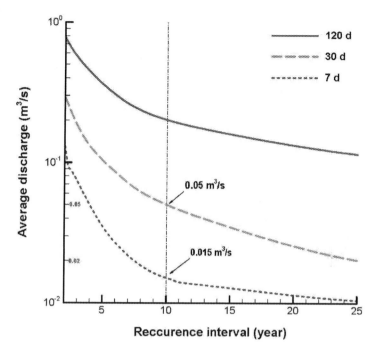

Figure 1.3 An example of low flow frequency curve of a typical creek.

1.2 WATER ENVIRONMENT MODELING IN WATER POLLUTION CONTROL

1.2.1 Evolution of water pollution control

In 1849, there was a cholera outbreak that killed many people in London, England. At that time, an epidemiological investigation was conducted by Dr. John Snow. The results of the investigation pointed out that the source of cholera was a well, which was also the source of drinking water for the nearby residents. Subsequently, the well contaminated with cholera bacteria was abandoned, and the cholera epidemic gradually subsided. This incident has called the attention of modern society to water pollution prevention and control, and the world recognizes that the success of water pollution prevention and control is to determine and cut off the pollution source (ReVelle and ReVelle, 1988).

The focus of the early water pollution prevention and control practice was to maintain public health and to curb the spread of aquatic infectious diseases such as cholera and typhoid fever. There are many types of bacteria for different aquatic infectious diseases, and the probability of detecting a specific bacterium in the water environment is low. Therefore, it is time consuming to determine the types and quantities of bacteria in a certain water sample. To overcome this difficulty, *Escherichia coli* is selected as the indicator organism for water pollution control as it is ubiquitous in the water environment and easy to measure. Under normal circumstances, *E. coli* is harmless to the human body, but high level of *E. coli* in water indicates that there is a high risk of bacterial contamination. Untreated domestic sewage contains approximately 1,000,000 *E. coli* per 100 mL, and the drinking water should not contain more than 200 *E. coli* per 100 mL. Since the 1920s, disinfection has become an important procedure in water supply and sewage treatment, which has effectively controlled various types of waterborne infectious diseases. Chlorination is the most common method for water disinfection, while ozone and ultraviolet rays are used sometimes. When the treated sewage is released back into the river, coliform bacteria will be further reduced due to the natural dilution and decay effect (die-off).

In the 1920s, the US Public Health Services conducted a large-scale survey of water pollution in the Ohio River. The investigation found that the discharge of urban wastewater without proper treatment into the river would not only cause public health problems such as bacteria breeding but also cause the river water to become dirty and smelly due to the decrease of dissolved oxygen (DO), which consequently leads to massive death of fish and ecological deterioration. Since then, ecological conservation has been incorporated into the water pollution prevention and control practice together with public health protection.

DO in river water is necessary for the growth and reproduction of various aquatic animals and plants. The amount of DO in unpolluted river water remains close to saturation. The Ohio River Water Pollution Investigation Report pointed out that when urban sewage containing a large amount of organic waste is discharged into the river, microorganisms in the river will decompose the organic waste and consume DO. As the level of the DO decreases, high-grade fishes such as trout in the river gradually disappear and are replaced by pollution-resistant silver carp and grass carp. When the DO in the water is below 2.0 g/m^3, most fishes cannot survive. The reduction of DO also destroys the natural ecology in river. When the DO is completely lost, the river enters an anaerobic state. At this stage, the river turns black and emits a foul smell. Because the oxygen consumption of organic waste in the water has a huge impact on water quality, the term biochemical oxygen demand (BOD) was proposed to represent the amount of organic waste, and it is still in use nowadays. BOD is the amount of DO consumed when organic waste is decomposed in water.

The rate of decomposition of organic waste in water is called the rate of deoxygenation. When the level of DO in the river is lower than saturation, oxygen in the atmosphere will enter the water. The rate of oxygen replenishment is called the rate of reaeration. When the reaeration rate of a river is less than the rate of deoxygenation, various pollution states will appear. Each river has a certain degree of self-purification ability or waste assimilative capacity. The early environmental engineering literature defined self-purification factor of a river simply as the ratio of reaeration rate to the deoxygenation rate.

The initial impression of river pollution to people is that the original clear water becomes turbid after a large amount of sediment flows in. Therefore, the quantity of suspended solids has always been an important parameter in water pollution control. Organic waste exists in the polluted water in the form of suspended solids and dissolved substances. Early sewage treatment was the primary treatment based on sedimentation of suspended solids. The primary treatment can remove most of the suspended solids in the sewage, as well as up to 50% of the BOD.

In 1972, the US Congress enacted the Water Pollution Prevention and Control Act, commonly known as the 1972 Clean Water Act. Prior to this, most of the urban sewage and industrial wastewater in the United States were discharged into rivers or other receiving water bodies without any treatment or only after primary treatment, and so the pollution load of these water bodies far exceeded their self-purification capacity, resulting in serious pollution of the water environment. The 1972 Water Purification Program contained the following three regulations related to the discharge of urban sewage and industrial wastewater.

1.2.1.1   Minimum treatment requirements for sewage discharge

Under the 1972 Clean Water Act, municipal sewage treatment plants are required to meet secondary treatment standards. According to the definition of the USEPA, secondary sewage treatment must remove at least 85% of the suspended solids and BOD in the sewage. In untreated urban sewage, there are about 200 mg/L of suspended solids and 200 mg/L of 5-d BOD (BOD_5). Therefore, after secondary treatment, the suspended solids and BOD_5 in the sewage should be less than 30 mg/L. As for industrial wastewater, the 1972 Clean Water Act has similar treatment requirements, which is called Best Available Treatment. Because the wastewater produced by various industries is very different, there are different optimal treatment requirements for different industrial wastewater, and the details are determined by the USEPA. After 25 years of hard work, by 1989, all major cities and small towns in the United States had built and operated sewage treatment plants with secondary or higher-level treatment.

1.2.1.2   Permit system for urban sewage and industrial wastewater discharge

According to the Clean Water Act of 1972, a wastewater discharge permit (National Pollutants Discharge Elimination System [NPDES]) must be obtained before urban sewage and industrial wastewater can be discharged into rivers. The permit sets the requirements for the water quality and regulations for the relevant on-site monitoring. The water quality requirement is determined by water quality modeling of the receiving river. If the modeling results indicate that the water quality of the river after receiving wastewater discharge from secondary treatment can meet or exceed its water quality standard, the secondary treatment will be the water quality requirement on the permit. Otherwise, further water quality modeling will be performed to calculate the allowable effluent water quality requirement. Section 1.2.2 introduces how to apply water quality models to determine the requirements of the discharge water quality. Detailed discussions are presented in Chapter 6.

1.2.1.3   Federal government funding for the construction of wastewater treatment plants

The United States 1972 Clean Water Act requires the federal government to support 75% of the construction cost of municipal wastewater treatment plants (WWTPs). When applying for construction funding from the USEPA, the wastewater facility planning report must be submitted. An important part of this report is to analyze the self-purification capacity of the receiving rivers using water quality modeling and to calculate the degree of sewage treatment required.

After 25 years' effort, WWTPs have been built across the United States, and a significant progress has been made in the prevention and control of water pollution. However, the national water quality report issued by the USEPA (1998) revealed that the water quality of more than 40% of rivers and other water bodies still failed to meet the standard. The report further pointed out that the main reasons were the neglect of nonpoint source pollution and the impact of nontraditional pollutants on water quality, including nutrients and toxic substances. Since the 1990s, the focus of water pollution control in the United States has gradually shifted from point source to nonpoint source pollution control. Meanwhile, the research and development of the prevention and control methods for toxic substances and new pollutants (emerging pollutants) has been strengthened.

1.2.2   Water environment modeling fundamentals: (1) surface water

When the wastewater discharge after secondary treatment still causes pollution to a river, according to the 1972 Clean Water Act, higher-level of wastewater treatment is necessary. Any wastewater treatment that can achieve a pollutant removal rate higher than the secondary treatment standard is called advanced wastewater treatment. Advanced wastewater treatment methods including membrane filtration and activated carbon adsorption are usually expensive. Therefore, the pollutant removal rate required for advanced sewage treatment must be calculated correctly using water quality modeling (Figure 1.4). Since the implementation of the Clean Water Act 1972, the water quality model has become an indispensable analysis tool in water environment management, and the derivation and application of the water quality model has become a new and hot topic in environmental engineering.

The self-purification capability of a river depends on the combined effect of its hydrodynamic characteristics and natural reaction mechanism. The river water quality model combines these two based on the law of mass conservation. In the early stage, the Streeter–Phelps model was used to simulate the self-purification capacity of rivers and determine the requirements for pollutant removal rate. In this simple and practical river water quality model, the river hydraulic mechanism is represented by the average flow velocity, and the natural reaction mechanism is represented by the oxygen-consuming reaction and the river reaeration. After 1972, many complex river water quality models were derived and applied to the planning and design of large-scale wastewater treatment facilities (refer to Chapter 6).

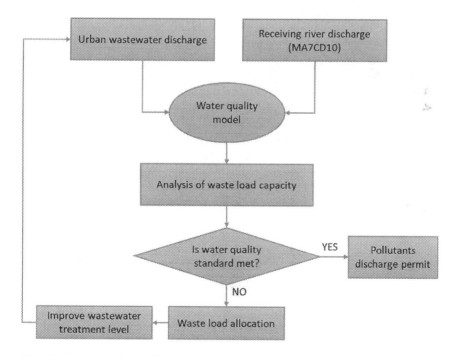

Figure 1.4 Water quality modeling in river water quality management.

For prevention and control of point source pollution, the pollution load released into a receiving river is also the design discharge of a WWTPs. The river water quality model can be used to simulate the self-purification capacity of the river. An initially planned amount of pollution discharge is set as the input data for the water quality model, and the resultant river water quality is calculated and compared with the water quality standard. If the river water quality calculated by the model is better than the water quality standard, the initially planned discharge amount will become the discharge amount in the wastewater discharge permit and can be used to design a WWTP. Otherwise, it is necessary to further reduce the pollution discharge and repeat the model analysis (Figure 1.4).

Sewage disinfection was originally regarded as a necessary part of secondary sewage treatment, usually the last procedure before sewage is discharged. In later studies, it was found that many rivers have high self-purification capability against *E. coli*, which can be rapidly reduced through natural dilution and die-off. Therefore, in most cases, sewage disinfection for *E. coli* was not necessary. Moreover, if the sewage was disinfected by chlorine treatment, carcinogens such as trihalomethanes would be produced, and the residual chlorine in water would also have a negative impact on river ecology. Therefore, after considering these research results, the USEPA decided not to include sewage disinfection as part of secondary treatment (USEPA, 1973). Meanwhile, the USEPA also required that each sewage treatment plant should conduct river water quality modeling to evaluate the self-purification capability of the river against *E. coli* and determine whether sewage disinfection is necessary.

Urban sewage and industrial wastewater are usually discharged into rivers or other receiving water bodies through discharge pipes at fixed locations. This type of water pollution is usually called point source pollution. The discharge of point source pollution is relatively stable and does not vary with time significantly. Therefore, pollution problems usually occur during the dry period of the receiving river. The organic mass in the river sediment consumes the DO during decomposition, which can be regarded as nonpoint source pollution and affect the river water quality in the dry season. However, this impact is limited and not included in the early river water quality models.

Generally, nonpoint source pollution is caused by pollutants carried in the surface runoff induced by a heavy rainfall in the watershed. Therefore, when modeling the point and nonpoint source pollution at the same time, the low flow discharge MA7CD10 is inadequate, instead the unsteady flow of the receiving river after heavy rain should be used. The USEPA proposed the concept of total maximum daily load (TMDL) as the basis for the prevention and control of nonpoint source pollution. The maximum daily pollution load is the maximum amount of pollution that a river can receive everyday while maintaining its water quality standard, which can be expressed by the following formula:

$$\text{TMDL} = \left(L_w\right)_p + \left(L_w\right)_{np} + \text{MOS} \tag{1.1}$$

In Eqn. (1.1), $\left(L_w\right)_p$ is the point source waste load, $\left(L_w\right)_{np}$ is the nonpoint source waste load, and MOS is the margin of safety, the safety factor considering other uncertainties.

The point source waste load in Eqn. (1.1) can be calculated using the water quality model introduced earlier, and it is assumed that the receiving river is in the MA7CD10 low flow state. In order to calculate the nonpoint source waste load in Eqn. (1.1), water quantity and quality models for watersheds are utilized to calculate runoff and waste amount generated under heavy rain condition, and then, the river water quality model is used to analyze the waste load allowed while still maintaining the water quality standard. As nonpoint source pollution is caused by runoff induced by rainstorms, the flow of receiving rivers should not be the MA7CD10 low flow. Therefore, Eqn. (1.1) actually gives a conservative estimation of the maximal allowable pollutant load. The water quality models related to the prevention and control of nonpoint source pollution will be discussed in Chapter 5.

One of the biggest water pollution problems caused by nonpoint source pollution in the receiving water is eutrophication and algae bloom. These problems usually occur in lakes, reservoirs, and low-speed sections of rivers. If the water bodies are unaffected by human activities, the growth of algae and other aquatic plants will be limited by nutrient supply. Under such condition, although other growth conditions are available, due to the limited supply of specific nutrient, aquatic plants will not grow excessively, and a good ecological balance can be maintained. This specific nutrient is called a limiting nutrient. Phosphates are usually the limiting nutrients in most inland lakes and rivers, while nitrogen or nitrates are often the limiting nutrients in lakes and rivers in coastal areas.

Lake Erie on the border of the United States and Canada once suffered from eutrophication. Reducing the limiting nutrients discharged into the lake is the main solution for this problem. The lake water quality models can be used to calculate the amount of limiting nutrients allowed to discharge. In the early stage, the simple Vollenweider model (Chapra, 1997) was adopted in the water quality management of Lake Erie and other water bodies suffering from eutrophication. This model regards the lake as a CSTR. The degree of eutrophication and improvement methods can be evaluated by modeling the natural reaction and average retention time of the limiting nutrients (nitrogen or phosphorus) in the lake. Detailed discussion of CSTR and other simple models are discussed in Chapter 6.

More comprehensive lake water quality modeling can be conducted on the basis of two-dimensional or three-dimensional time-dependent models. These complex models involve numerical solutions using finite difference method (FDM) or finite element method (FEM; Young, 2002; Kuo and Yang, 2002). Water quality models are also employed in the pollution control after accidental spills of hazardous materials into water environment. A well-known example is the application of water quality models in the remediation of polychlorinated biphenyl (PCB) pollution in the Hudson River in the eastern United States (Thomann and St. John, 2007; Connolly et al., 2000).

1.2.3 Water environment modeling fundamentals: (2) groundwater

One of the most frequently mentioned cases of groundwater pollution in environmental protection movement is the Love Canal incident in Buffalo, New York, USA. During the decade of 1950s, the Hooker Chemical Plastic Corporation used the deserted canal bed as a dump site and buried more than 20,000 metric tons of toxic chemicals. The company then sold this 16-acre plot of land to the local district at a low price, where a primary school and hundreds of houses were built. After 1970, the high occurrence of abnormal infants and other health-related problems in that area attracted the attention of local health authorities, and a detailed investigation was conducted. The investigation revealed that this serious environmental pollution incident was caused by the toxic waste dumped by Hooker Corporation entering the household basement and the primary school playground via underground water. This incident aroused the public concern of groundwater pollution in the United States, and subsequently in 1980, the United States Congress enacted a bill commonly known as the Superfund (USEPA, 1992). The Superfund requires all chemical contaminated sites like Love Canal to be investigated and identified across the United States. Since then, environment restoration of these contaminated sites has been financially supported by the Superfund, which has been in turn shared by all chemical companies in the United States.

In the environment restoration process of Love Canal, the groundwater quality modeling played a very important role. The Hooker Corporation once owned four illegal landfills near Buffalo. Figure 1.5 shows the modeling results of one of the landfills called the S-Area, where 63,000 metric tons of toxic chemicals were dumped. The biggest concern at the landfill is the invasion of nonaqueous-phase liquids (NALPs) into the fractured bedrock aquifer. The simulation results indicated that without restoration measures, NALPs would infiltrate into the bedrock aquifer after heavy rains. However,

Before

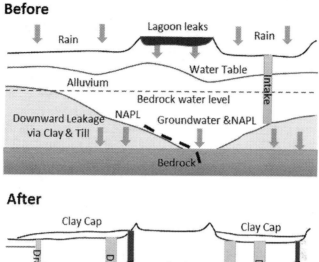

After

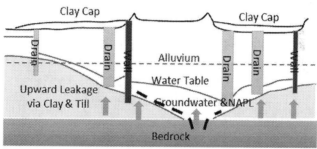

Figure 1.5 Application of groundwater quality modeling in pollution control in Buffalo, New York.

if restoration measures were adopted, such as drainage pipes, barrier walls, and clay covering, the pollutant would change its flow direction, moving away from the bedrock aquifer (Cohen et al., 1987).

Groundwater quality model is also an effective analysis tool for water resource management. Groundwater is a good source of water because its quantity and quality are more stable than rivers and other surface waters. More than 20% of the public water supply in the United States comes from groundwater, and some cities use groundwater as the main source of public water supply.

Groundwater is the only source of water for nearly a million people in Honolulu in the Hawaiian Islands. In Honolulu, to ensure the stability of water supply, it is necessary to understand and calculate the sustainable yield of groundwater, which is the maximum amount of groundwater that can be pumped without affecting the water supply capability of the groundwater layers. The surface soil and the volcanic rock below the islands of Hawaii have high water permeability. After rainfall reaches the ground, a large portion of water infiltrates into the ground and becomes groundwater, while only a small part becomes surface water. Naturally, the groundwater of Honolulu slowly flows from the volcanic rock to the surrounding ocean. Before reaching the Pacific Ocean, the groundwater is buffed by the caprock layer in the coastal zone. The caprock is formed by the deposition of decayed animals and plants over hundreds of thousands of years, and its water permeability is very small. When the groundwater flow enters the caprock, the flow rate drops sharply, and the water level rises. Therefore, the caprock layer functions a natural hydraulic dam, behind which a huge natural underground reservoir is formed. Hawaiian groundwater, also known as basal aquifer, is different from continental groundwater, which generally exists in sand and gravel. The fresh water in a basal aquifer is floating above the sea water (Figure 1.6). Therefore, the thickness of the freshwater layer can be determined according to the vertical salinity distribution data obtained from the observation well. The largest natural underground reservoir in Honolulu Island is the Pearl Harbor Reservoir. The freshwater layer in this reservoir is 300 m thick. Currently, the Honolulu Water Company's deep wells draw approximately 640,000 metric tons of

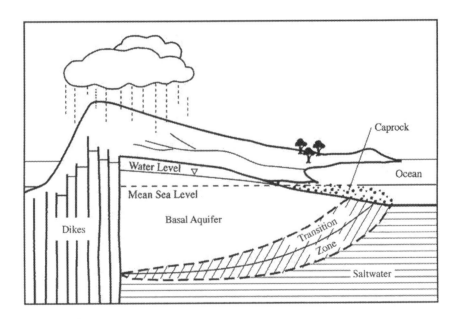

Figure 1.6 Hydrogeological map of the groundwater layer in Pearl Harbor, Hawaii.

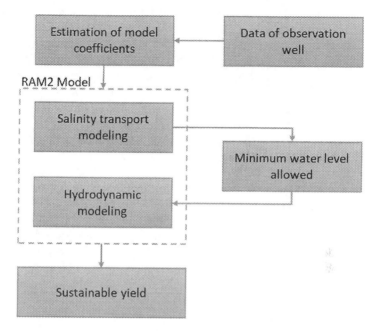

Figure 1.7 Analysis procedure of groundwater quality model RAM2 model.

water from various groundwater layers every day, among which the Pearl Harbor underground reservoir has the largest pumping capacity, about 400,000 metric tons per day.

Overexploitation of groundwater will lead to the decline of groundwater level and seawater intrusion. Researchers have established a variety of complex two-dimensional and three-dimensional groundwater flow and transport models. However, before using these models to calculate the sustainable yield, it is necessary to estimate correctly a series of coefficients required as input. Based on the current understanding of hydrogeology and groundwater flow, a robust analytical groundwater flow and salinity transport model (RAM2 [Robust Analytical Model 2]) was developed. The coefficients of RAM2 model can be correctly estimated on the basis of the records of deep observation wells (Liu, 2008). Nowadays, RAM2 modeling provides the basis for groundwater management in Honolulu city (Figure 1.7).

1.3 WATER QUALITY MODEL APPLICATIONS

1.3.1 Uncertainty in modeling analysis

The results of water quality modeling are highly dependent on the model coefficients. Model calibration or verification is the process of matching the calculated results with the measured data by adjusting the model coefficients within a reasonable range.

Photosynthesis and respiration of aquatic plants in rivers can cause significant diurnal variations in the DO. Therefore, model calibration based on the measured DO data without indicating when the data were measured can lead to great uncertainty (Liu and Fok, 1983). Figure 1.8 shows comparison between calculated and measured DO in a river. The model employed is the modified Streeter–Phelps model, a one-dimensional steady-state river model, which can calculate only the variation of the DO with distance. Moreover, the field survey only records where the DO data were recorded and ignores the specific time when the measurement is performed. Under such conditions, model calibration by adjusting model coefficients, such as the oxygen consumption coefficient and the reaeration coefficient, does not necessarily reflect the true reaction mechanism of the river.

The uncertainty in water quality modeling also comes from the estimation of model coefficients. Take the estimation of reaeration coefficient as an example. The reaeration effect is strongly dependent on the turbulence intensity of water flow. Currently, empirical formulas are commonly used to calculate reaeration coefficient in river water quality modeling. These formulas are derived by measuring the change of DO level under reaeration effect in open channels, with several hydraulic parameters representing turbulence intensity, such as water depth, velocity, and hydraulic gradient of the riverbed. The commonly used formulas for reaeration coefficient are O'Connor-Dobbins formula (O'Connor and Dobbins, 1956) and Churchill formula (Churchill et al., 1962). The two formulas can produce quite

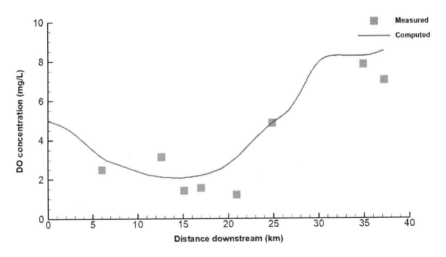

Figure 1.8 Model correction of DO without consideration of diurnal variation effect in a river.

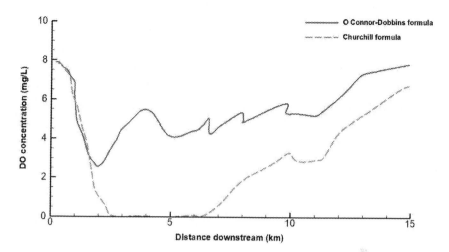

Figure 1.9 Modeling results of river's DO with different reaeration formulas.

different results when they are applied to the water quality modeling of natural rivers (Rathbun, 1977).

For example, water quality modeling is used to determine whether the DO standard for a river can be met, and the results are presented in Figure 1.9. It can be seen in Figure 1.9 that the simulated DO concentration in the river is above the water quality standard when the O'Connor-Dobbins formula is used to estimate the reaeration coefficient. However, if the Churchill formula is used, most of the river sections cannot meet the water quality standard. Therefore, using different empirical formulas for reaeration coefficient will result in difference modeling results. In practice, the tracer method can be used to measure the actual reaeration coefficient and reduce modeling uncertainty.

Uncertainty can also be produced in the process of numerical solution, which can be explained using numerical dispersion. The common numerical method for solving unsteady river water quality model is the FDM. When the FDM is used for the advection term, numerical dispersion would normally occur, which in turn increases the uncertainty of calculating the actual river dispersion process.

1.3.2 Basic procedures for water environment model application

Basic procedures must be followed to ensure that the water quality modeling provides reliable results for water pollution prevention and water resource management. Figure 1.10 shows a complete procedure for river

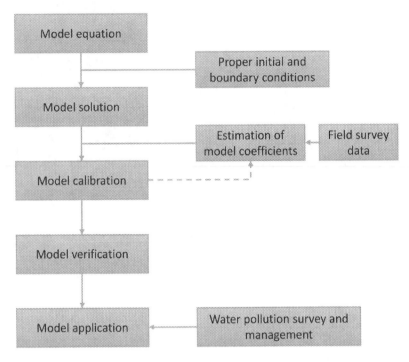

Figure 1.10 Analysis procedure of river water quality modeling.

water quality modeling. Although the discussion in this section is focused on river water quality modeling, the principles are also applicable to water quality modeling for other surface water or groundwater.

1.3.2.1   Mathematical formulation

The selection of different water quality models should be determined by the purpose of water quality modeling and data available to verify the model. Whether a water quality modeling is successful should be measured by whether the modeling purpose is achieved. If a simple water quality model can fulfill the purpose of analysis, it should be used with priority. The simple model is derived from the complex ones by introducing reasonable simplification assumptions. As a result, the simple models require less computation resources and can be completed in a shorter period of time. Another factor in model selection is the model coefficients. The simple model involves a smaller number of model coefficients, and thus, the estimation and calibration of model coefficients would cost less efforts. When a complex model is required, it is advised to start from the simple one, as the results by the simple model can be used to guide the formulation and application of the complex models.

1.3.2.2 Mathematical solution

Water quality models normally consist of differential equations and initial/boundary conditions. The water quality models generally need to be solved using numerical methods, except for the simple models that have analytical solutions for certain problems. In numerical solution, the river is divided into a grid system, and the partition of the grids determines whether the simulation is effective and accurate.

1.3.2.3 Model calibration

Model calibration is a process of adjusting model coefficients within reasonable ranges so that the computed water quality results, such as DO, are close to the measured data. The model coefficients can be estimated on the basis of the data obtained from laboratory measurements and field survey. The generalized environmental conditions in laboratory and the actual conditions in the field can be quite different, which will affect the reaction mechanism represented by the model coefficients. Therefore, it is advised to include field survey as part of the water quality modeling process.

1.3.2.4 Model verification

After model calibration, another field survey should be conducted under different hydraulic and environmental conditions. The modeling results should be compared to the new field data. If the calculated and measured results agree well, the model is considered verified. Otherwise, the model coefficients must be further adjusted.

1.3.2.5 Model application

The verified river water quality models can be applied to simulation of self-purification capability of a river and to predict water quality when pollutants are discharged into the river. Therefore, water quality models are an effective tool for water quality management and water pollution prevention. In practice, water quality models can be used to calculate the maximum pollution load allowed for a river, which is the maximum amount that a river can receive under the designed flow and environmental conditions while still maintaining its water quality standard.

1.4 PROSPECTS FOR WATER ENVIRONMENT MODELING

1.4.1 Integration of traditional and linear systems modeling approaches

A contaminated water body can be treated as a linear system. According to the system theory, the self-purification capability of the water body can be represented by the impulse response function of the linear system, and the pollution discharge can be represented by the input function of the linear system. The impulse response function and the input function can be estimated separately and independently. The change of water quality in the water caused by pollution can be represented by the linear system output function, which can be calculated by a simple convolution integration of the impulse response function and the input function. Therefore, the traditional water quality models consist of a set of differential equations and associated boundaries and initial conditions, while the linear water quality models are made up of integral equations. The conjunctive application of the two types of models is a new trend for water quality modeling (Liu, 1986).

1.4.2 Modeling with modern information technology

A large amount of data is needed in water quality modeling. The data management is one of the keys to achieve the modeling goals and to clearly present modeling results to the public and decision makers. The combination of the water quality models and modern information technology is illustrated through USEPA's watershed information management platform BASINS (refer to Chapter 5).

1.4.3 Modeling with modern biological technology

The ultimate goal of water pollution control is to protect aquatic environmental ecology and public health, which should be developed along with modern biotechnology. At present, water quality measuring instruments are commonly manufactured on the basis of physical and chemical principles, while biological sensors under development can provide more useful data for water quality modeling. In addition, combining aquatic plant growth mechanisms with river hydrodynamics can help to further understand wave–plant interaction and to improve the accuracy of estimating the water quality parameters such as reaeration coefficients (Figure 1.11; Liu et al., 2015).

1.4.4 Modeling the fate and transport of emerging contaminates in water environment

Over the past two decades, the detection and reporting of emerging contaminants (ECs) in the aquatic environment have raised concern for water quality managers and environmental regulators. ECs such as pharmaceuticals and personal care products, endocrine disrupting chemicals, and pesticides are now frequently detected globally in surface water and have been linked to adverse ecological effects (Luo et al., 2014). There is evidence that effluent discharge from WWTPs is a major pathway for the introduction of ECs to the aquatic environment (Blair et al., 2013; Wilkinson et al., 2017;

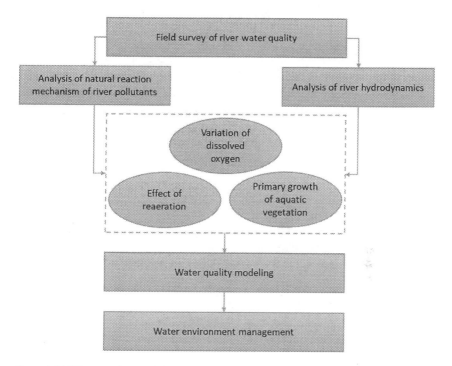

Figure 1.11 Water quality model combined with modern biological technology.

Li et al., 2019). Hence, a better understanding of their fate and the removal of ECs in WWTPs and the environment would facilitate the management and control of these environmental contaminants.

ECs are commonly present in waters at trace levels ranging from ng/L to µg/L, which makes their detection and quantification a challenge (Luo et al., 2014). Modeling is therefore regarded as a useful approach to predict their fate during wastewater treatment and in the receiving environment (Pomiès et al., 2013). Fugacity-based multimedia fate models describe the processes controlling chemical behavior in environmental media by developing and applying mathematical statements of chemical fate (Mackay, 2001). These models could comprehensively illustrate the distribution and concentrations of chemicals in various 'compartments' within a system (Mackay and Paterson, 1991). They can also provide quantitative analysis of the emission sources, transport, and transfer routes, as well as sinks of ECs within the systems under investigation (MacLeod et al., 2010). An advantage of the fugacity approach is that the use of fugacity, instead of concentration, makes it easy to use in mass balance calculations to describe the behavior of chemicals released into the environment and to simulate how chemicals are directionally transported between several compartments (Mackay and Paterson, 1991; Su et al., 2019; Wang et al., 2020).

EXERCISES

1. According to Figure 1.2, when a city requires 5.6 m³/s of stable water supply, which river type should be used, river type 1 or river type 2. Why?
2. Is there any reservoir near your resident area? Collect and study the literature about the quantity and quality of the nearest major reservoir in your location, and please write a report to address the following issues:

   (a) Summary of the hydrological features of the reservoir and its importance to public water supply for your area.
   (b) Discuss the water quality modeling of the reservoir.
   (c) How to use the water quality modeling as an effective analysis tool in the planning of sustainable operation of the reservoir.

REFERENCES

Blair, B.D., Crago, J.P., Hedman, C.J., Treguer, R.J., Magruder, C., Royer, L.S. and Klaper, R.D. (2013). Evaluation of a model for the removal of pharmaceuticals, personal care products, and hormones from wastewater. *Science of the Total Environment*, 444, pp. 515–521.

Chapra, S.C. (1997). *Surface Water-Quality Modeling.* McGraw-Hill, Boston, MA.

Churchill, M.A., Elmore, H.L. and Buckingham, R.A. (1962). The prediction of stream aeration rates. *Journal of the Sanitary Engineering Division ASCE*, 88(SA4), pp. 1–46.

Cohen, R.M., Rabold, R.R., Faust, C.R., Rumbaugh III, J.O. and Bridge, J.R. (1987). Investigation and hydraulic containment of chemical migration: Four landfills in Niagara Falls. *Civil Engineering Practice*, Spring, pp. 33–58.

Connolly, J.P., Zahakos, H.A., Benaman, J., Ziegler, C.K., Rhea, J.R. and Russell, K. (2000). A model of PCB fate in the Upper Hudson River. *Environmental Science & Technology*, 34(19), pp. 4076–4087.

International Energy Agency. (2015). *Key World Energy Statistics.* International Energy Agency (IEA), Paris, France.

Kuo, J.T. and Yang, M.D. (2002). Water quality modeling in reservoirs. *Environmental Fluid Mechanics: Theories and Applications*, pp. 377–420.

Li, L., Zou, D., Xiao, Z., Zeng, X., Zhang, L., Jiang, L. and Liu, F. (2019). Biochar as a sorbent for emerging contaminants enables improvements in waste management and sustainable resource use. *Journal of Cleaner Production*, 210, pp. 1324–1342.

Liu, C.C.K. (1986). Surface water quality analysis. Chapter 1 in Wang, L. and Pereira, N.C. (eds.), *Handbook of Environmental Engineering.* The Humana Press, Clifton, NJ, pp. 1–59.

Liu, C.C.K. (2008). RAM2 modeling and the determination of sustainable yield of Hawaii basal aquifer. *Project Report PR-2008–06, Water Resources Research Center*. University of Hawaii, Manoa, Honolulu, HI.

Liu, C.C.K. and Fok, Y.S. (1983). Stream waste assimilative capacity analysis using reaeration coefficients measured by tracer techniques. *JAWRA Journal of the American Water Resources Association*, 19(3), pp. 439–445.

Liu, C.C.K., Lin, P., Xiao, H., Zhang, X., Liu, B. and Zhang, Y. (2015). Evaluation of stream reaeration capacity based on field water quality survey data. *Proceedings of the 17th Mainland-Taiwan Environmental Protection Conference*. Kunming, China, December 6–11, 2015.

Luo, Y., Guo, W., Ngo, H.H., Nghiem, L.D., Hai, F.I., Zhang, J., Liang, S. and Wang, X.C. (2014). A review on the occurrence of micropollutants in the aquatic environment and their fate and removal during wastewater treatment. *Science of the Total Environment*, 473, pp. 619–641.

Mackay, D. (2001). *Multimedia Environmental Models: The Fugacity Approach*. CRC Press, Boca Raton, FL.

Mackay, D. and Paterson, S. (1991). Evaluating the multimedia fate of organic chemicals: A level III fugacity model. *Environmental Science & Technology*, 25(3), pp. 427–436.

MacLeod, M., Scheringer, M., McKone, T.E. and Hungerbuhler, K. (2010). The State of Multimedia Mass-balance Modeling in Environmental Science and Decision-making. *Environmetal Science & Technology*, 44(22), pp. 8360–8364.

O'Connor, D.J. and Dobbins, W.E. (1956). The mechanism of reaeration in natural stream. *Journal of the Sanitary Engineering Division ASCE*, 82(6), pp. 1–30.

Overman, M. (1969). *Water: Solutions to a Problem of Supply and Demand*. Doubleday & Company Inc, New York.

Pomiès, M., Choubert, J.-M., Wisniewski, C. and Coquery, M. (2013). Modelling of micropollutant removal in biological wastewater treatments: A review. *Science of the Total Environment*, 443, pp. 733–748.

Rathbun, R.E. (1977). Reaeration coefficients of streams – State of the art. *Journal of the Hydraulics Division, ASCE*, 103(HY4), pp. 409–424.

ReVelle, P. and ReVelle, C. (1988). *The Environment – Issues and Choices for Society*. Jones and Barlett Publishers, Boston.

Su, C., Zhang, H., Cridge, C. and Liang, R. (2019). A review of multimedia transport and fate models for chemicals: Principles, features and applicability. *Science of the Total Environment*, 668, pp. 881–892.

Thomann, R.V. and St. John, J.P. (2007). The fate of PCBs in the Hudson River ecosystem. *Annals of the New York Academy of Sciences*, 320(1), pp. 610–629.

U.S. Environmental Protection Agency. (1973). *Guidelines for Developing or Revising Water Quality Standards Under the Federal Water Pollution Control Act Amendment of 1972*. Water Planning Division, Washington, DC.

U.S. Environmental Protection Agency. (1992). *CERCLA/SUPERFUND Orientation Manual*. USEPA Technology Innovation Office, EPA/542/R-92/005. Washington, DC.

U.S. Environmental Protection Agency. (1998). *Water Pollution Control – 25 Years of Progress and Challenges for the New Millennium*. EPA833-F-98-003. Washington, DC.

Wang, Y., Fan, L., Khan, S.J. and Roddick, F.A. (2020). Fugacity modelling of the fate of micropollutants in aqueous systems – Uncertainty and sensitivity issues. *Science of the Total Environment*, 699, p. 134249.

Wilkinson, J., Hooda, P.S., Barker, J., Barton, S. and Swinden, J. (2017). Occurrence, fate and transformation of emerging contaminants in water: An overarching review of the field. *Environmental Pollution*, 231, pp. 954–970.

Young, D.L. (2002). Finite element analysis of stratified lake hydrodynamics. Chapter in Shen, H.H., Cheng, A.H.D., Wang, K.H., Teng, M.H. and Liu, C.C.K. (eds.), *Environmental Fluid Mechanics, Theories and Applications*. American Society of Civil Engineers, Reston, VA, pp. 339–376.

Chapter 2

Environmental hydraulics and modeling

2.1 MASS CONSERVATION PRINCIPLE

Mass conservation principle states that the mass of an object or collection of objects never changes, no matter how the constituent parts rearrange themselves. The principle was first discovered and applied to chemical reactions by Antoine Laurent Lavoisier in 1785 (Kotz et al., 2009). Since then, it has been widely used in the fields of chemistry, fluid flow, and mass transfer (Daily and Harleman, 1972).

2.1.1 Mass continuity equation

As for mass transfer in a fluid, mass conservation principle can be defined as 'the rate of mass change in a control volume equals the summation of mass flux across its surface plus the rate of reaction'. Control volume is a fixed volume in the space, and it does not change with time. Figure 2.1 shows a transporting substance flows through a control volume V. The surface area of this control volume is S, the concentration of the transporting substance is C, and the reaction rate is r.

According to the mass conservation principle, the rate of change of the mass of a substance in a control volume $\dfrac{\partial M}{\partial t}$ equals its mass flux $-J(t)$ plus the rate of reaction r. For simplicity, a conservative substance with its concentration of C is first considered, for which the reaction rate r is zero, and the mass conservation principle takes the following form:

$$\frac{\partial M}{\partial t} = -J(t) \tag{2.1}$$

As the total mass of the substance in the control volume M can be expressed by a volume integral, the rate of change of the mass in a control volume can be expressed as follows:

$$\frac{\partial M}{\partial t} = \frac{\partial}{\partial t} \iiint C dV = \iiint \frac{\partial C}{\partial t} dV \tag{2.2}$$

DOI: 10.1201/9781003008491-2

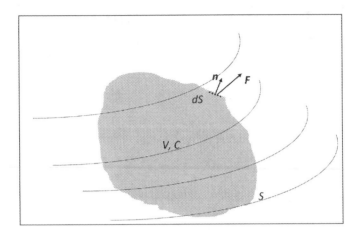

Figure 2.1 Control volume in a fluid flow.

Note that the control volume, a fixed volume in the space, does not change with time, and thus, the integration variables in Eqn. (2.2) are constants and the sequence of integration and differentiation can be interchanged.

As shown in Figure 2.1, F is the mass flux out of a small surface area ds. The total mass flux out of the entire control volume can be determined by a surface integration:

$$J(t) = \iint F \cdot n \, ds \qquad (2.3)$$

where n is a unit vector perpendicular to a differential area ds.

Applying divergence theorem, the surface integration in Eqn. (2.3) can be transformed to a volume integration (Sokolnikoff and Redheffer, 1966):

$$J(t) = \iint F \cdot n \, ds = \iiint \nabla \cdot F \, dV \qquad (2.4)$$

where $\nabla \cdot F$ is a dot product or $\nabla \cdot F = \dfrac{\partial F_x}{\partial x} + \dfrac{\partial F_y}{\partial y} + \dfrac{\partial F_z}{\partial z}$.

Substitute Eqns. (2.2) and (2.4) into Eqn. (2.1) to get:

$$\iiint \left(\frac{\partial C}{\partial t} + \nabla \cdot F \right) dV = 0 \qquad (2.5)$$

Equation (2.5) is valid for any portion of the control volume. Therefore, the integrand must be zero everywhere, and Eqn. (2.5) may be rewritten as follows:

$$\frac{\partial C}{\partial t} = -\nabla \cdot F \qquad (2.6)$$

Equation (2.6) is the mathematical formulation of mass conservation principle of a conservative transporting substance in a fluid and is often referred to as the mass continuity equation. For a nonconservative transporting substance, Eqn. (2.6) takes the following form:

$$\frac{\partial C}{\partial t} = -\nabla \cdot \mathbf{F} + r \tag{2.6a}$$

2.1.2   Mass transport equation

Generally, mass flux of a transporting substance in water environment is caused by advective mass flux ($\mathbf{F}_{adv}$) and diffusive mass flux ($\mathbf{F}_{dif}$):

$$\mathbf{F} = \mathbf{F}_{adv} + \mathbf{F}_{dif} \tag{2.7}$$

The advective mass flux of transporting substances in a water environment is produced by the mean velocity of the moving water and can be calculated by the following equation:

$$\mathbf{F}_{adv} = C\mathbf{q} \tag{2.8}$$

where $\mathbf{q} = u\mathbf{i} + v\mathbf{j} + w\mathbf{k}$ is a velocity vector of the water flow.

The diffusive mass flux of transporting substances in a water environment can be described by Fick's law of molecular diffusion that was formulated by Adolf Fick (1829–1901), analogous to the Fourier law of heat conduction (see Csanady, 1973). The Fourier law of heat conduction states that the conduction velocity of heat energy in a material depends on the temperature gradient in the material and the heat conduction coefficient. Similarly, Fick's law of molecular diffusion states that diffusive mass flux depends on the gradient of molecular concentration of a transport substance in a fluid and the molecular diffusion coefficient. Mathematically, Fick's law can be expressed as follows:

$$\mathbf{F}_{dif} = -D_m \nabla C \tag{2.9}$$

where D_m is the molecular diffusion coefficient and ∇C is the molecular concentration gradient of a transporting substance in a fluid. The values of molecular diffusion coefficient of a few common substances are listed in Table 2.1.

Substituting Eqns. (2.8) and (2.9) into Eqn. (2.7), the total mass flux can be expressed as follows:

$$\mathbf{F} = \mathbf{q}C - D_m \nabla C \tag{2.10}$$

With Eqn. (2.10), Eqn. (2.6) becomes:

$$\frac{\partial C}{\partial t} + \nabla \cdot (\mathbf{q}C) - \nabla \cdot (D_m \nabla C) = 0 \tag{2.11}$$

Table 2.1 Molecular diffusion coefficients of a few common substances

Diffusing Molecule	Diffusion Coefficient (cm²/s)
CO_2 in air	0.137
CO_2 in water	1.8×10^{-5}
Water vapor in air	0.220
Hydrocarbons in air	$0.05 \sim 0.08$
N_2 in water	2.0×10^{-5}
NaCl in water	1.24×10^{-5}

Source: Rohsenow and Choi (1961).

Equation (2.11) is the advection–molecular diffusion equation of a conservative transporting substance in water environment. Two assumptions are often introduced in the application of Eqn. (2.11): (1) the fluid is incompressible, $\nabla \cdot q = 0$, then $\nabla \cdot (qC) = q \cdot \nabla C$; and (2) molecular diffusion coefficient D_m is a constant. With these assumptions, Eqn. (2.11) can be simplified to:

$$\frac{\partial C}{\partial t} + q \cdot \nabla C - D_m \nabla^2 C = 0 \tag{2.12}$$

where $\nabla^2 = \dfrac{\partial^2}{\partial x^2} + \dfrac{\partial^2}{\partial y^2} + \dfrac{\partial^2}{\partial z^2}$

Expand Eqn. (2.12) into Cartesian coordinates to give:

$$\frac{\partial C}{\partial t} + u\frac{\partial C}{\partial x} + v\frac{\partial C}{\partial y} + w\frac{\partial C}{\partial z} - D_x\frac{\partial^2 C}{\partial x^2} - D_y\frac{\partial^2 C}{\partial y^2} - D_z\frac{\partial^2 C}{\partial z^2} = 0 \tag{2.13}$$

2.2   MOLECULAR DIFFUSION

There are about 10^{26} water molecules in one cubic centimeter of water. While moving randomly, these molecules collide with one another. After other substances enter a water body, they are subject to rapid collision by randomly moving water molecules and mix outward from the entering point. This mixing phenomenon is called diffusion if the sizes of these substances are in the same order of magnitude as that of water molecule, while the mixing is called Brownian motion if these substances are much larger than water molecule (Csanady, 1973). In environmental fluid mechanics, both diffusion and Brownian motion are often referred to as molecular diffusion, and both molecules and particles of dissolved and suspended substances are referred to as diffusing molecules.

Under normal circumstances, molecular diffusion is not an important mechanism of pollutant transport in water environment as the molecular diffusion coefficients of common substances in water environment are very small (Table 2.1). However, as will be introduced later in this book, all physically based water environment models can be formulated on the basis of the fundamental advection–diffusion equation, simply by replacing the molecular diffusion coefficient in Eqn. (2.13) by a much larger turbulent diffusion coefficient or a shear flow dispersion coefficient. In addition, the analytical solution of the basic molecular diffusion equation forms the basis for solving many practical water environment problems. Therefore, the basic molecular diffusion equation – its derivation, analytical solution, and physical significance – will be discussed in detail in this section.

2.2.1   The basic molecular diffusion model and its analytical solution

For a one-dimensional advection–diffusion analysis, Eqn. (2.13) can be simplified to:

$$\frac{\partial C}{\partial t} + u\frac{\partial C}{\partial x} - D_x\frac{\partial^2 C}{\partial x^2} = 0 \tag{2.14}$$

If a water is still or velocity $u = 0$, by setting D_x to be the molecular diffusion coefficient D_m, Eqn. (2.14) can be further simplified to:

$$\frac{\partial C}{\partial t} = D_m\frac{\partial^2 C}{\partial x^2} \tag{2.15}$$

For a one-dimensional analysis, it is important to clarify the units of substance concentration and mass. The concentration of a substance in a one-dimensional water system can be expressed in term of g/m^3 or g/m. When g/m^3 is used as concentration unit, the corresponding mass unit is g/m^2. Conversely, when g/m is used as concentration unit, the corresponding mass unit is g.

There are several methods to derive analytical solutions of Eqn. (2.15) under varying initial/boundary conditions. The most fundamental solution of Eqn. (2.15) corresponds to the molecular diffusion of a substance after it is introduced at a very high concentration at the origin or an impulse input. The corresponding initial/boundary conditions can be defined mathematically as follows:

$$C(x,0) = M\delta(0) \tag{2.16}$$

where $\delta(0)$ is the Dirac delta function $\delta(t)$ when $t = 0$ (see Figure 2.2 for illustration).

Equations (2.15) and (2.16) together are called the basic molecular diffusion model. The basic molecular diffusion model consists of two major components: (1) the governing equation, or Eqn. (2.15), and (2) initial/boundary conditions, or Eqn. (2.16).

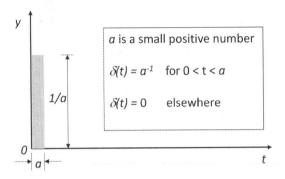

Figure 2.2 Dirac delta function $\delta(t)$.

Strictly speaking, $\delta(t)$ does not satisfy rigorous definition of an ordinary function; rather, it can be better recognized by its action on an ordinary function. The significance of the Dirac delta function and its application in engineering mathematics are best realized by its Laplace transform.

As shown later, Laplace transformation is to transform an ordinary function in a time domain to a Laplace domain.

$$F(s) = \int_0^\infty f(t)e^{-st}\,dt \tag{2.17}$$

where $f(t)$ is an ordinary function.

Based on Eqn. (2.17), the Laplace transformation of the Dirac delta function is expressed as follows:

$$L[\delta_a(t)] = \int_0^\infty a^{-1}e^{-st}\,dt = (sa)^{-1}\left(1 - e^{-st}\right) \tag{2.18}$$

Expanding $(1-e^{-st})$ in Taylor series for e^{-sa}, it can be shown that $L[\delta_a(t)] = 1$ when a is a very small positive number. In other words, the Dirac delta function has a value of 1 in the Laplace domain. This is an important relationship in engineering mathematics and can be applied here to derive the analytical solution for the basic molecular diffusion model.

Take a Laplace transformation of the left-hand side of Eqn. (2.15) and then integrating by parts, we obtain:

$$\int_0^\infty e^{-st}\frac{\partial C}{\partial t}\,dt = \left[Ce^{-st}\right]_0^\infty + s\int_0^\infty Ce^{-st}\,dt = sF \tag{2.19}$$

In deriving Eqn. (2.19), the initial condition $C = 0$ at $t = 0$ was introduced, and when $t = \infty$, $e^{-st} = 0$. Therefore, $\left[Ce^{-st}\right]_0^\infty = 0$.

Take a Laplace transformation of the right-hand side of Eqn. (2.15) and change the sequence of integration and differentiation to get:

$$D_m \int_0^\infty e^{-st} \frac{\partial^2 C}{\partial x^2} dt = D_m \frac{\partial^2}{\partial x^2} \int_0^\infty C e^{-st} dt = D_m \frac{d^2 F}{dx^2} \tag{2.20}$$

Therefore, with Laplace transformation, the governing PDE of the basic molecular diffusion model becomes the following ordinary differential equation:

$$D_m \frac{\partial^2 F}{\partial x^2} = sF \tag{2.21}$$

As the Laplace transform of the Dirac delta function is 1, Eqn. (2.16) in the Laplace domain takes the following form:

$$F(x = 0) = M \tag{2.22}$$

With a simple integration, the ordinary differential equation (Eqn. (2.21)) with the boundary conditions defined by Eqn. (2.22) can be solved, and the solution is expressed in a Laplace domain:

$$F = \frac{M}{\sqrt{4D_m}} \frac{e^{-\sqrt{\frac{s}{D_m}}x}}{\sqrt{\frac{s}{D_m}}} \tag{2.23}$$

By an inverse Laplace transform, the aforementioned solution in a Laplace domain can be transformed back to the time domain. By using the Laplace transform table, Eqn. (2.23) can be readily transformed back to a time domain (Crank, 1975):

$$C(x,t) = \frac{M}{\sqrt{4\pi D_m t}} e^{-\left(\frac{x^2}{4D_m t}\right)} \tag{2.24}$$

Equation (2.24) is the analytical solution of the basic molecular diffusion model. This solution can also be derived by using the method of dimensional analysis (Fischer et al., 1979) and by other methods (Crank, 1975).

The solution of the basic molecular diffusion equation or Eqn. (2.24) is plotted in Figure 2.3. As shown in Figure 2.3, after transporting substance of a very high concentration at time $t = 0$ at the origin, a plume of the transporting substance will be formed and spread outward. The transporting plume at any time would take the form of a normal or Gaussian distribution. As time increases, the mass concentration in the center of the plume reduces and the base of the plume expands.

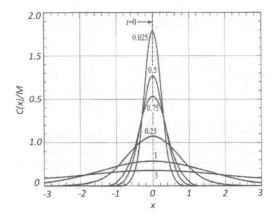

Figure 2.3 Molecular diffusion and normal distribution.

By definition, the first three moments of a normal distribution take the following form (Fischer et al., 1979):

$$\text{Zeroth moment } M_0 = \int_{-\infty}^{\infty} C(x,t)dx \tag{2.25}$$

$$\text{First moment } M_1 = \int_{-\infty}^{\infty} xC(x,t)dx \tag{2.26}$$

$$\text{Second moment } M_2 = \int_{-\infty}^{\infty} x^2 C(x,t)dx \tag{2.27}$$

The mean μ and the variance σ^2 of a one-dimensional normal distribution can then be defined by these moments:

$$\mu = \frac{\int_{-\infty}^{\infty} xC(x,t)dx}{\int_{-\infty}^{\infty} C(x,t)dx} = \frac{M_1}{M_0} \tag{2.28}$$

$$\sigma^2 = \frac{\int_{-\infty}^{\infty} (x-\mu)^2 C(x,t)dx}{M_0} = \left(\frac{M_1}{M_0}\right) - \mu^2 \tag{2.29}$$

Substituting Eqn. (2.24) into aforementioned definitions of moments, the mean and variance of a diffusing plume can be calculated by carrying out integrations, and the results are $\mu = 0$ and $\sigma^2 = 2D_m t$.

The standard deviation σ that is the square root of the variance is also used to denote the spatial distribution of a diffusing plume. Since σ is a function of time, it can be used in Eqn. (2.24) to replace time t as an independent variable. As a result, the changes of a diffusing plume as a function of (x, t) can be expressed as a function of (x, σ) as follows:

$$C(x,t) = \frac{M}{\sigma\sqrt{2\pi}} e^{-\left(\frac{x^2}{2\sigma^2}\right)}$$

(2.30)

According to Eqn. (2.24) or Eqn. (2.30), $C = 0$ when x approaches $\pm\infty$. Therefore, the substance concentration in a diffusing plume would not entirely disappear until at a distance infinitely far from the source, or the width of a diffusing plume is infinitely large. In practice, a finite width of a diffusing plume needs to be defined. The most popular definition of this width is four times of the standard deviation or 4σ. It can be shown that, by carrying out the integration in Eqn. (2.30), the total mass of a diffusing plume with a width of 4σ would include approximately 95% of the total mass.

Because $\sigma^2 = 2D_m t$, the rate of change of the variance with respect to time is:

$$\frac{d\sigma^2}{dt} = 2D_m$$

(2.31)

Equation (2.31) indicates that the variance of a diffusing plume changes with time at a rate of $2D_m$. In other words, the diffusion coefficient can be determined when the rate of change of the variance of a diffusion or dispersion plume is known. This principle has been applied to estimate the longitudinal dispersion coefficient of a river reach by using the field tracer method in surface water quality modeling (see Section 7.2.3) and in ground-water modeling (see Section 9.2.5).

The following example shows that the Laplace transformation can also be applied for solving the basic molecular diffusion model with varying boundary/initial conditions.

Example 2.1 Analytical solution of the basic molecular diffusion model with a time-varying input function.

Problem: The initial concentration in a one-dimensional diffusion system is zero everywhere. At $t = 0$, the concentration is suddenly raised to C_0 at $x = 0$ and then kept in that level there. Find the concentration distribution in the system $C(x,t)$.

Solution: The model of this problem can be expressed mathematically as follows:

Governing equation (2.15):

$$\frac{\partial C(x,t)}{\partial t} = D_m \frac{\partial^2 C}{\partial x^2}$$

Initial/boundary conditions:

$$C(x,0) = 0, \quad -\infty < x < \infty$$
$$C(0,t) = C_0, \quad t > 0$$

(2.32)

Take the Laplace transformation of unknown variable $C(x,t)$,

$$F = \int_0^x C(x,t)e^{-st}dt$$

And then take the Laplace transform of the first and second derivatives of unknown variable C, $\dfrac{\partial C}{\partial t}$, and $\dfrac{\partial^2 C}{\partial x^2}$:

$$\int_0^x e^{-st}\frac{\partial C}{\partial t}dt = \left[Ce^{-st}\right]_0^x + s\int_0^x Ce^{-st}dt = sF \qquad (2.33)$$

And

$$\int_0^x e^{-st}\frac{\partial^2 C}{\partial x^2}dt = \frac{\partial^2}{\partial x^2}\int_0^x Ce^{-st}dt = \frac{\partial^2 F}{\partial x^2} \qquad (2.33a)$$

After the Laplace transform, the governing equation changes to be an ordinary differential equation:

$$D_m\frac{d^2 F}{dx^2} = sF \qquad (2.34)$$

The general solution of this ordinary differential equation takes the following form:

$$F = e^{-mx} \qquad (2.35)$$

Substituting Eqn. (2.35) into Eqn. (2.34), Eqn. (2.34) is further simplified to an algebraic equation:

$$\left(D_m m^2 - s\right)e^{-mx} = 0 \qquad (2.36)$$

The solution of m in Eqn. (2.36) is expressed as follows:

$$m = \pm\sqrt{\frac{s}{D_m}} \qquad (2.37)$$

Insert Eqn. (2.37) into Eqn. (2.35) to get:

$$F = Ae^{-\sqrt{\frac{s}{D_m}}x} + Be^{\sqrt{\frac{s}{D_m}}x} \qquad (2.38)$$

When $x \to \infty$, $Be^{\sqrt{\frac{s}{D_m}}x} \to \infty$; therefore, the coefficient B in Eqn. (2.38) must be zero and thus:

$$F = Ae^{-\sqrt{\frac{s}{D_m}}x} \qquad (2.39)$$

The constant A can be determined by using the boundary condition $C(0, t) = C_0$, which in the Laplace domain is $\dfrac{C_0}{s}$:

$$F = \frac{C_0}{s} = Ae^{-\sqrt{\frac{s}{D_m}}0} = A \tag{2.40}$$

Therefore, after Laplace transformation, the solution of a basic molecular diffusion model with a time-varying input function is given as follows:

$$F = \frac{C_0}{s} e^{-\sqrt{\frac{s}{D_m}}x} \tag{2.41}$$

By using the Laplace transform table, the solution of Eqn. (2.41) can be transformed back into the time domain,

$$C(x,t) = C_0 \left[1 - \mathrm{erf}\left(\frac{x}{\sqrt{4D_m t}} \right) \right] \tag{2.42}$$

where $\mathrm{erf}(x)$ is called error function, which is defined as $\mathrm{erf}(x) = \dfrac{2}{\sqrt{\pi}} \displaystyle\int_0^x e^{-m^2} dm$. Equation (2.42) is often put in the following form:

$$C(x,t) = C_0 \left[\mathrm{erfc}\left(\frac{x}{\sqrt{4D_m t}} \right) \right] \tag{2.42a}$$

where $\mathrm{erfc}(x) = 1 - \mathrm{erf}(x)$ is called the complementary error function. The solution or Eqn. (2.42) is illustrated by Figure 2.4.

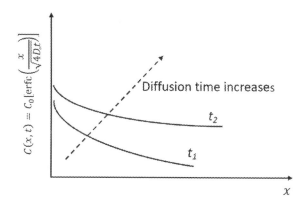

Figure 2.4 Concentration distribution in a one-dimensional diffusion system receiving a constant input $C = C_0$ at $x = 0$.

Comments

1. The analytical solution of a one-dimensional diffusion system that receives a time-variable input at the boundary $x = 0$ can be derived by using the Laplace transformation. Equation (2.42) is the solution when the input function is a constant $C(0,t) = C_0$.

2. The solution for a one-dimensional diffusion equation receiving a time-varying input can also be derived by the method of dimensional analysis. Refer to Fischer et al. (1979, pp. 43–45).

With constant molecular diffusion coefficients, the analytical solution for a one-dimensional diffusion system can be readily extended to two-dimensional and three-dimensional diffusion systems as shown in Example 2.2.

Example 2.2 Analytical solution of a two-dimensional diffusion equation receiving a pulse input.

Problem: Derive the analytical solution of a two-dimensional diffusion system with constant diffusion coefficients receiving an impulse mass input at the origin.

Solution: Two-dimensional diffusion model with constant diffusion coefficients takes the following form:

$$\frac{\partial C}{\partial t} = D_x \frac{\partial^2 C}{\partial x^2} + D_y \frac{\partial^2 C}{\partial y^2} \tag{2.43}$$

For molecular diffusion, $D_x = D_y = D_m$. In environmental mixing problems, however, turbulent diffusion or dispersion coefficients in different directions are most likely different.

After a pulse mass input is introduced to this diffusion system, the initial condition can be expressed as follows:

$$C(x,y,0) = M\sigma(x)\sigma(y) \tag{2.44}$$

Applying the product rule, $C(x, y, t)$ can be expressed as follows:

$$C(x,\ y,\ t) = C_1(x,\ t)\ C_2(y,\ t) \tag{2.45}$$

where C_1 is not a function of y, and C_2 is not a function of x, thus:

$$\frac{\partial(C_1 C_2)}{\partial t} = C_1 \frac{\partial C_2}{\partial t} + C_2 \frac{\partial C_1}{\partial t} = D_x C_2 \frac{\partial^2 C_1}{\partial x^2} + D_y C_1 \frac{\partial^2 C_2}{\partial y^2} \tag{2.46}$$

Equation (2.46) can be re-arranged as follows:

$$C_2 \left[\frac{\partial C_1}{\partial t} - D_x \frac{\partial^2 C_1}{\partial x^2} \right] + C_1 \left[\frac{\partial C_2}{\partial t} - D_y \frac{\partial^2 C_2}{\partial y^2} \right] = 0 \qquad (2.47)$$

Equation (2.47) is true only if items inside both parentheses equal to zero. In other words, C_1 and C_2 must satisfy one-dimensional model individually. In addition, to satisfy the mass conservation principle requires $\iint C dx dy = M$. Thus, the analytical solution of a two-dimension diffusion model receiving a pulse mass input at the origin is given as follows:

$$C = C_1 C_2 = \frac{M}{4\pi t \sqrt{D_x D_y}} e^{-\frac{x^2}{4D_x t} - \frac{y^2}{4D_y t}} \qquad (2.48)$$

Comments

Similarly, the analytical solution to a three-dimension diffusion model with constant diffusion coefficients and receiving an impulse mass input at the origin is given as follows:

$$C = C_1 C_2 C_3 = \frac{M}{(4\pi t)^{2/3} \sqrt{D_x D_y D_z}} e^{-\frac{x^2}{4D_x t} - \frac{y^2}{4D_y t} - \frac{z^2}{4D_z t}} \qquad (2.48a)$$

2.2.2 The random walk model of molecular diffusion

It has been shown in the last section that diffusive flux of a transporting substance in water depends on substance concentration gradient and diffusion coefficient and that the basic molecular diffusion model can be formulated on the basis of the mass conservation principle and Fick's Law of molecular diffusion. Furthermore, the plume created by molecular diffusion after the transporting substance entering a water as an impulsive input possesses a normal distribution. It will be shown in this section that molecular diffusion can also be simulated statistically by using a random walk model.

The random walk model can be introduced simply by considering random walk of a particle that makes a unit step length along a straight line, and the probability of either a forward or a backward step is exactly 50%. After a particle being released at the origin $x = 0$ has walked N steps, this particle could reach any of the points: $-N, -N + 1, -N + 2 \ldots -1, 0, +1 \ldots N - 2, N - 1, N$. The probability of the particles to fall into a specific point m between $-N$ and N is $p(m, N)$, which can be determined by the following formula (Csanady, 1973):

$$p(m, N) = \sqrt{\frac{2}{\pi N}} e^{-\left(\frac{m^2}{2N}\right)} \qquad (2.49)$$

In a continuous one-dimensional system, the location of the specific point m can be expressed as follows:

$$m = \frac{x}{l_w} \tag{2.50}$$

where l_w is the length of one random walk step.

By the definition of random walk, the distance of two adjacent particles is $2l_w$, or there are $\Delta x/2l_w$ particles in an incremental distance Δx. Consequently, the probability of one particle in Δx is $p(m,N)\dfrac{\Delta x}{2l_w}$. If the particles of a total mass M are released at $x = 0$, the mass in the incremental distance Δx should be:

$$\Delta M = Mp(m,N)\frac{\Delta x}{2l_w} \tag{2.51}$$

and the distribution of the diffusing particles in Δx is expressed as follows:

$$C = \frac{\Delta M}{\Delta x} = Mp(m,N) = \frac{M}{2l}\sqrt{\frac{2}{\pi N}}e^{-\left(\frac{x^2}{2Nl_w^2}\right)} \tag{2.52}$$

Letting the release time of each particle to be $1/\theta$, the diffusing time t and the total number of the released particles N can be redefined as $t = \dfrac{N}{\theta}$ or $N = t\theta$. By defining the diffusion coefficient D_m as $D_m = \dfrac{1}{2}\theta l_w^2$, Eqn. (2.52) becomes:

$$c(x,t) = \frac{M}{\sqrt{4\pi D_m t}}e^{-\left(\frac{x^2}{4D_m t}\right)} \tag{2.53}$$

Note that Eqns. (2.53) and (2.24) are identical. Thus, the following observation can be made:

> After a substance enters in a water body, dissolved or suspended particles of this substance would be agitated by moving water molecules and move randomly. A pulse input of this substance will spread out and form a normal distribution. This result can be simulated by a basic molecular diffusion model which is formulated based on Fick's law of diffusion. This result can also be simulated statistically by a random walk model.

2.2.3 One-dimensional advection–diffusion analysis

The basic molecular diffusion of a conservative substance in a still water was discussed in the last two sections. In reality, environment water is rarely still but moves at varying velocities. For a moving water, two kinds of flow are defined on the basis of the relative magnitude of flow velocity and fluid viscosity. When the influence of water viscosity on flow is much larger than water particle velocity, the flow is called laminar flow; otherwise, the flow is called turbulent flow. In fluid mechanics, relative magnitude of water particle velocity and water viscosity is measured by the Reynolds number (Massey, 1998). As shown in Example 2.3, the mixing of a conservative substance in a laminar flow can be readily simulated by extending the basic molecular diffusion equation. The mixing of a nonconservative substance in turbulent flow is much more complicated and will be discussed in the next section.

Example 2.3 One-dimensional advection–molecular diffusion.

Problem: Derive the analytical solution of a one-dimensional advection–molecular diffusion system receiving a pulse mass input.

Solution: Because the basic diffusion equation or Eqn. (2.15) is a linear equation, the governing equation of a one-dimensional advection–diffusion model can be formulated by adding an advection term to the basic diffusion equation:

$$\frac{\partial C}{\partial t} = -u\frac{\partial C}{\partial x} + D_m\frac{\partial^2 C}{\partial x^2} \tag{2.54}$$

A new coordinate system (ξ, τ), $\xi = x - ut$, $\tau = t$ is defined. Relative to this new system, Eqn. (2.54) can be converted to the following form:

$$\frac{\partial C}{\partial \tau} = D_m\frac{\partial^2 C}{\partial \xi^2} \tag{2.55}$$

The new coordinate system is essentially a moving coordinate system with a constant velocity u. A pulse mass input can be expressed similar to Eqn. (2.16) as $C(\xi, 0) = M\delta(0)$.

Because the governing equation and initial/boundary condition have the same form as the basic molecular diffusion model, the analytical solution of a one-dimensional advection–diffusion system in a moving coordinate can be derived simply by replacing variables (x,t) in Eqn. (2.24) with (ξ,τ) or

$$C(\xi,\tau) = \frac{M}{\sqrt{4\pi D_m\tau}}e^{-\frac{\xi^2}{4D_m\tau}} \tag{2.56}$$

Converting this solution back to the original (x,t) coordinate, the analytical solution of a one-dimensional advection–diffusion system is expressed as follows:

$$C(x,t) = \frac{M}{\sqrt{4\pi D_m t}}e^{-\frac{(x-ut)^2}{4D_m t}} \tag{2.57}$$

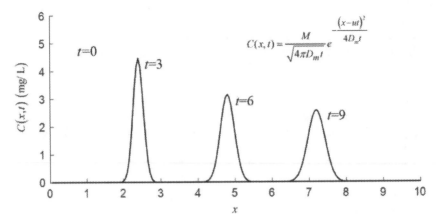

Figure 2.5 Substance transports in a one-dimensional advection–diffusion system receiving a pulse input.

Comments

After a conservative substance at a high concentration is released at the origin of a one-dimensional advection–diffusion system, a substance plume is formed and moves downstream at the flow velocity u. The arrival of the plume center at any downstream location depends on the water travel time, or $t = x/u$. When this plume moves downstream, the substance concentration in this plume remains to be of the normal distribution, with its peak decreasing and the base expanding (Figure 2.5).

2.3 TURBULENT DIFFUSION AND SHEAR FLOW DISPERSION

2.3.1 Turbulent diffusion and advection–turbulent diffusion equation

Laminar flow or the unidirectional multilayer movement of a moving fluid occurs only when the viscous force of the fluid is much larger than its inertial force. When the inertial force of a moving fluid is much larger than its viscous force, the flow will become turbulent in which individual fluid particles undergo erratic movement in three dimensions. Although the size of particles that take erratic movements in a turbulent flow are small, they are much larger than individual molecules. Most of the natural water bodies such as rivers, lakes, and coastal waters are under turbulent flow conditions, and erratic movements of particles would enhance the mixing of dissolved and suspended substances in these water bodies, in addition to molecular diffusion.

Water velocity at a fixed point in turbulent flow can be expressed as follows:

$$u(t) = \bar{u} + u'(t) \tag{2.58}$$

where $\bar{u}$ is the average water velocity at the point and $u'(t)$ is the fluctuating velocity (Figure 2.6).

As shown in Figure 2.6, a water particle moves in turbulent flow as a superposition of an average velocity and a small-scale irregular motion. A rigorous mathematical analysis of turbulent flow is still not possible, and hence, the mixing of dissolved and suspended substances in turbulent flow must be investigated by statistical methods.

The statistical analysis of the mixing of dissolved and suspended substances was first conducted by G.I. Taylor with a series of experiments (Taylor, 1921; Taylor, 1954). In these experiments, a large amount of tracer particles was released repeatedly into a stationary and homogeneous turbulent flow. The particle movement in terms of displacement and velocity at a fixed point was observed, and the ensemble average value was calculated. On the basis of experimental results, Taylor found that the magnitude and distribution of tracer plumes formed in the flow can be described by the ensemble average of the displacement of tracer particles. Furthermore, the ensemble average of tracer particle displacement and the ensemble average of fluctuating velocity are related by the following equation:

$$\left\langle X^2(t) \right\rangle = \int_0^t \int_0^t \left\langle u'(\tau_1) u'(\tau_2) \right\rangle d\tau_1 d\tau_2 \tag{2.59}$$

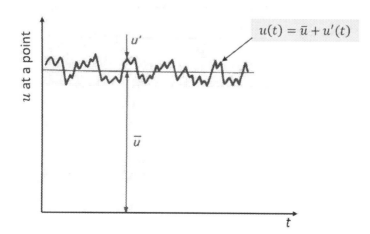

Figure 2.6 Water velocity at a fixed point in a turbulent flow.

where $\langle X^2(t) \rangle$ and $\langle u'(\tau_1) u'(\tau_2) \rangle$ are ensemble averages of the square of particle displacement and fluctuating velocity, respectively.

For a stationary flow, the ensemble average of fluctuating velocity is a function of $(\tau_2 - \tau_1)$ only. Therefore, a correlation coefficient $R_x(\tau_2 - \tau_1)$ can be defined as follows:

$$R_x(\tau_2 - \tau_1) = \frac{\langle u'(\tau_1) u'(\tau_2) \rangle}{\langle u^2 \rangle} \tag{2.60}$$

where $\langle u^2 \rangle = \langle u(0) u(0) \rangle$ is the characteristic constant for a particular turbulent flow and is called turbulence intensity.

With Eqn. (2.60), Eqn. (2.59) can be expressed as follows:

$$\langle X^2(t) \rangle = \langle u^2 \rangle \int_0^t \int_0^t R_x(\tau_2 - \tau_1) d\tau_1 d\tau_2 \tag{2.61}$$

To further simplify Eqn. (2.61), two new variables S and τ are defined as follows: $s = \tau_2 - \tau_1$; $\tau = \dfrac{\tau_2 + \tau_1}{2}$. Introducing these two variables into Eqn. (2.61) and performing integration with respect to τ, we get:

$$\langle X^2(t) \rangle = 2 \langle u^2 \rangle \int_0^t (t - s) R_x(s) ds \tag{2.62}$$

Based on Eqn. (2.62), the following two limiting values of the ensemble average of the displacement of tracer particles exist:

1. Shortly after a tracer substance is released into a turbulent flow, velocities of tracer particles remain relatively constant, and thus, its movement is not entirely random. Therefore, its correlation coefficient approaches unity, and Eqn. (2.62) becomes:

$$\langle X^2(t) \rangle \rightarrow 2 \langle u^2 \rangle t^2 \tag{2.63}$$

2. Long after a tracer substance is released into a turbulent flow, its movement is entirely random, and the correlation coefficient approaches zero. Assuming $\int_0^x R_x(s) ds$ and $\int_0^x s R_x(s) ds$ still exist, we have:

$$\langle X^2(t) \rangle \rightarrow 2 \langle u^2 \rangle T_x t + \text{constant} \tag{2.64}$$

where $T_x = \int_0^x R_x(s) ds$ is called the Lagrangian time scale.

The differentiation of Eqn. (2.64) with respect to time t yields:

$$\frac{d\langle X^2(t)\rangle}{dt} = 2\langle u^2\rangle T_x \qquad (2.65)$$

Both turbulence intensity $<u^2>=<u(0)u(0)>$ and Lagrangian time scale $T_x = \int_0^\infty R_x(s)ds$ are characteristic constants for a particular turbulent flow. Taylor replaced the right-hand side of Eqn. (2.65) by a single constant ε:

$$\frac{d\langle X^2\rangle}{dt} = 2\varepsilon \qquad (2.66)$$

Equation (2.66) is very similar to that of Eqn. (2.31), which shows the rate of growth of the variance of a normally distributed plume produced by molecular diffusion. Comparing Eqn. (2.66) with Eqn. (2.31), it is found that, by defining ε as a turbulent diffusion coefficient, the effect of turbulent diffusion on particle mixing in a water body is similar to that of molecular diffusion. Therefore, Fick's law can be applied to turbulent diffusion as well as molecular diffusion. The rate of molecular diffusion depends on the gradient of molecular concentration and the molecular diffusion coefficient. Similarly, the rate of turbulent diffusion depends on the gradient of substance particle concentration and the turbulent diffusion coefficient. Replacing the molecular diffusion coefficient D_m with the turbulent diffusion coefficient ε, the mass flux produced by turbulent diffusion can be expressed as follows:

$$F_{dif} = -\varepsilon \nabla C \qquad (2.67)$$

Thus, turbulent advection–diffusion can be expressed in Cartesian coordinate by the following equation:

$$\frac{\partial C}{\partial t} + u\frac{\partial C}{\partial x} + v\frac{\partial C}{\partial y} + w\frac{\partial C}{\partial z} - D_x\frac{\partial^2 C}{\partial x^2} - D_y\frac{\partial^2 C}{\partial y^2}$$
$$-D_z\frac{\partial^2 C}{\partial z^2} - \varepsilon_x\frac{\partial^2 C}{\partial x^2} - \varepsilon_y\frac{\partial^2 C}{\partial y^2} - \varepsilon_z\frac{\partial^2 C}{\partial z^2} = 0 \qquad (2.68)$$

The values of turbulent diffusion coefficients ε_x, ε_y, and ε_z depend on the hydraulic properties of water environment, including the size of eddies. Therefore, turbulent diffusion coefficient is also called eddy diffusivity.

For most environmental waters, the value of turbulent diffusion coefficient is much larger than the value of molecular diffusion coefficient. For example, experimental results indicate that the values of turbulent diffusion coefficients in a river are approximately in the range of 10^0–10^6 cm²/s (Zison et al., 1978), while the molecular diffusion coefficient is only about

10^{-5}–10^{-1} cm²/s (see Table 2.1). In practical applications, therefore, the molecular diffusion term in Eqn. (2.68) can be ignored and a simplified advection–turbulent diffusion equation takes the following form:

$$\frac{\partial C}{\partial t} + u\frac{\partial C}{\partial x} + v\frac{\partial C}{\partial y} + w\frac{\partial C}{\partial z} - \varepsilon_x\frac{\partial^2 C}{\partial x^2} - \varepsilon_y\frac{\partial^2 C}{\partial y^2} - \varepsilon_z\frac{\partial^2 C}{\partial z^2} = 0 \qquad (2.69)$$

A one-dimensional advection–turbulent diffusion equation takes the following form:

$$\frac{\partial C}{\partial t} = -u\frac{\partial C}{\partial x} + \varepsilon_x\frac{\partial^2 C}{\partial x^2} \qquad (2.70)$$

It is important to note that Eqn. (2.69) was formulated assuming that the particle movement in a turbulent flow is entirely random. As discussed earlier in this section, the time required for the movement of tracer particles to become entirely random in a turbulent flow can be determined by Lagrangian time scale T_x. If the time of tracer particles residing in a turbulent flow is much shorter than Lagrangian time scale, Eqn. (2.69) cannot be applied to the simulation of the tracer particle transport.

Water environment models simulate transport in turbulent water bodies. Since the governing equation of a turbulent diffusion model has the same form as a molecular diffusion model, the analytical solution derived for simple molecular diffusion models can also be applied to simple turbulent diffusion models. The success of turbulent modeling would depend largely on how accurately the turbulent diffusion coefficient is determined.

The analytical solutions derived for simple basic molecular diffusion models usually assume that the diffusing system is infinitely large. In real water environment modeling, the solution depends on physical boundaries of the modeling system. The following example shows the application of the analytical solution of the basic molecular diffusion model for the transport simulation in a turbulent water body with two no-flow boundary conditions.

Example 2.4 The simulation of a one-dimensional turbulent diffusion system with solid boundaries.

Problem: Figure 2.7 shows an experimental water tank that has a width of 1 ft and a water depth of 8 ft. As the experiment starts at $t = 0$, 0.4 lb of a conservative tracer is injected at 5 ft from the left end as a source. Vertical oscillating screens are installed in the tank to generate flow turbulence with a measured turbulent diffusion coefficient of $\varepsilon = 0.05$ ft²/s.

1. Formulate the mathematical model of the turbulent transport of a tracer substance in the tank.
2. Calculate the peak tracer concentration at the left end and the time of its occurrence.

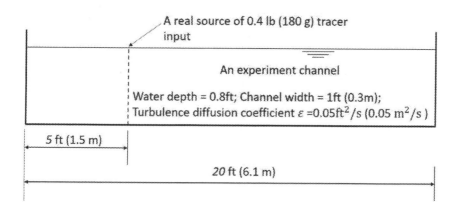

A real source of 0.4 lb (180 g) tracer input

An experiment channel

Water depth = 0.8ft; Channel width = 1ft (0.3m); Turbulence diffusion coefficient ε =0.05ft^2/s (0.05 m^2/s)

5 ft (1.5 m)

20 ft (6.1 m)

Figure 2.7 One-dimensional turbulent diffusion system with no-flow boundaries.

Solution: Because the tracer is distributed uniformly at any cross section and there is no advection, the transport system can be considered as a one-dimensional turbulent diffusion system with the following governing equation:

$$\frac{\partial C}{\partial t} = \varepsilon_x \frac{\partial^2 C}{\partial x^2} \qquad (2.71)$$

As shown in Figure 2.7, the left and right ends of the tank are located at $x = 0$ ft and $x = 20$ ft, and the no-flow system boundaries can be defined as follows:

$$\varepsilon \frac{\partial^2 C}{\partial x^2} = 0, \text{ at } x = 0 \text{ ft and } x = 20 \text{ ft} \qquad (2.72)$$

The initial conditions are as follows: $C(x,0) = 0$ except at $x = 5$, $C(5,0) = M$.

Note that if the concentration unit in this one-dimensional system is taken to be lb/ft^3, the corresponding mass unit should be lb/ft^2, that is, $M = 0.4/0.8 = 0.5$ lb/ft^2.

Equations (2.71) and (2.72) with aforementioned initial conditions constitute the model equations of this turbulent diffusion problem.

Because the model governing equation along with boundary and initial conditions are linear, it can be solved by the methods of images. According to the method of images, the effects of no-flow boundaries on tracer transport can be compensated by replacing these boundaries with images (Figure 2.8).

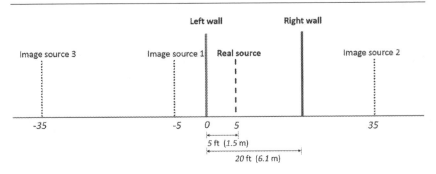

Figure 2.8 Solution of Example 2.4 by the method of images.

The solutions relative to each of the real and imaginary sources is Eqn. (2.24), which is the solution with a single source located in the center of an infinitely large transporting system. The desired solution is then the superposition of all these individual solutions.

$$C(0,t) = \frac{M}{\sqrt{4\pi\varepsilon t}} \sum_{n=1}^{\infty} e^{-\left(\frac{x_n^2}{4\varepsilon t}\right)} \tag{2.73}$$

where $x_1 = 5$ ft, $x_2 = -5$ ft, $x_3 = 35$ ft, $x_4 = -35$ ft, ...

In theory, the exact solution would be the superposition of infinite terms. However, in practical application, an acceptable approximate solution can be obtained by taking only the first two terms or

$$C(0,t) = \frac{M}{\sqrt{4\pi\varepsilon t}}\left[e^{-\left(\frac{5^2}{4\varepsilon t}\right)} + e^{-\left(\frac{(-5)^2}{4\varepsilon t}\right)}\right] = \frac{2M}{\sqrt{4\pi\varepsilon t}}\left[e^{-\left(\frac{25}{4\varepsilon t}\right)}\right] \tag{2.74}$$

The peak concentration can be calculated by taking the first derivative of Eqn. (2.74) and then setting the result to zero, $\dfrac{dC(0,t)}{dt} = 0$.

By trial-and-error method, the time of reaching the maximum concentration at the left end is $t_m = 250$ s. Introducing $t_m = 250$ s into Eqn. (2.74), the maximum tracer concentration is $C_m = 0.05$ lb/ft^3.

Comments

The analytical solution to the basic molecular diffusion model Eqn. (2.24) is derived for an infinitely large diffusion system. In case solid boundaries exist, the analytical solution is difficult to derive, and often the solution must be obtained by using numerical methods. Because the governing equation is linear, the method of images can be used to derive the approximate analytical solution, which is made up by an infinite series. In practical application, acceptable solution can be obtained by taking only the first two terms of this infinite series.

2.3.2   Shear flow dispersion and advection– dispersion equation

If the shear stress in a water flow causes a spatially fixed velocity distribution, the flow is called shear flow (Figure 2.9). Generally, water flows in natural and man-made open channels and pipes such as rivers, canals, and water supply pipes are all shear flow. In this section, additional mixing of tracer substance particles in a shear flow is discussed, first in a laminar shear flow and next in a turbulent shear flow.

Figure 2.9 shows a two-dimensional shear flow between two horizontal plates. The velocity distribution with respect to water depth can be expressed by the following equation:

$$u(y) = \bar{u} + u'(y) \tag{2.75}$$

where $\bar{u}$ is the average velocity and $u'(y)$ is the velocity deviation in space. The average velocity can be calculated by integration:

$$\bar{u} = \frac{1}{h} \int_0^h u\,dy \tag{2.76}$$

Similarly, variation of the concentration of tracer particles in vertical direction can be expressed as follows:

$$C(y) = \bar{C} + C'(y) \tag{2.77}$$

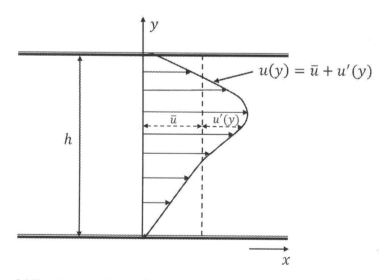

Figure 2.9 Two-dimensional shear flow.

where $\bar{C}$ is the average concentration and $C'(y)$ is the concentration deviation.

The average concentration can be calculated by an integration:

$$\bar{C} = \frac{1}{h} \int_0^h C \, dy \tag{2.78}$$

A moving coordinate system with two independent variables of (ξ, t) is introduced, where $\xi = x - \bar{u}t$ and $\tau = t$. In this coordinate system, the transport equation of a conservation substance in a two-dimensional laminar shear flow can be expressed as follows:

$$\frac{\partial \bar{C}}{\partial \tau} + \frac{\partial C'}{\partial \tau} + u' \frac{\partial \bar{C}}{\partial \xi} + u' \frac{\partial C'}{\partial \xi} = D_m \frac{\partial^2 C'}{\partial y^2} \tag{2.79}$$

Taylor (1954) conducted an experimental study on the tracer mixing in a laminar shear flow between two parallel plates. He observed that at any cross section, the tracer concentration remains constant at a particular water depth. He further observed that this constant concentration at a particular water depth is maintained by the balancing actions of longitudinal advection and lateral molecular diffusion, which can be expressed mathematically by the following equation:

$$u' \frac{\partial \bar{C}}{\partial \xi} = D_m \frac{\partial^2 C'}{\partial y^2} \tag{2.80}$$

Note that Eqn. (2.80) can also be derived by deleting three of the four terms on the right-hand side of Eqn. (2.79). The deletion is justified by considering the relative order of magnitude of the terms in Eqn. (2.79) (Fischer et al., 1979).

Two no-flow boundary conditions at the upper and lower plates can be defined as follows:

$$\frac{\partial C'}{\partial y} = 0 \text{ at } y = 0 \text{ and } y = h \tag{2.81}$$

where h is the water depth.

The solution to Eqn. (2.80) with the boundary conditions of Eqn. (2.81) is expressed as follows:

$$C'(y) = \frac{1}{D_m} \frac{\partial \bar{C}}{\partial \xi} \int_0^y \int_0^y u' \, dy \, dy \tag{2.82}$$

Relative to the moving coordinate, the mass flux of the transporting tracers as produced by shear flow F_{shear} can be expressed as follows:

$$F_{shear} = \int_0^b u'C'dy = \frac{1}{D_m}\frac{\partial \bar{C}}{\partial \xi}\int_0^b u'\int_0^y \int_0^y u'dydydy \qquad (2.83)$$

Equation (2.83) indicates that the mass flux of the transporting tracers as produced by shear flow is proportional to the substance concentration gradient or

$$F_{shear} = -bE\frac{\partial \bar{C}}{\partial \xi} \qquad (2.84)$$

where the constant E can be determined by comparing Eqns. (2.83) and (2.84):

$$E = \frac{1}{bD_m}\int_0^b u'\int_0^y \int_0^y u'dydydy \qquad (2.85)$$

Fick's law says that the rate of molecular diffusion depends on the concentration gradient of diffusing molecules and diffusion coefficient. For a one-dimensional flow, Fick's law can be expressed as follows:

$$F_x = -bD_m\frac{\partial C}{\partial x} \qquad (2.86)$$

Comparing Eqns. (2.84) and (2.86), it is clear that the constant E in Eqn. (2.84) plays the same role as the molecular diffusion coefficient D_m in Eqn. (2.86). Therefore, E is defined as the shear flow dispersion coefficient. Analogous to a one-dimensional molecular diffusion equation (Eqn. (2.15)), a one-dimensional shear flow dispersion equation can be formulated as follows:

$$\frac{\partial \bar{C}}{\partial \tau} = E\frac{\partial^2 \bar{C}}{\partial \xi^2} \qquad (2.87)$$

Returning to the Cartesian coordinate, the transport equation of a conservative tracer in a laminar shear flow takes the following form:

$$\frac{\partial \bar{C}}{\partial t} + u\frac{\partial \bar{C}}{\partial x} = E\frac{\partial^2 \bar{C}}{\partial x^2} \qquad (2.88)$$

Equation (2.88) can be extended to calculate the additional mixing of tracer particles in a turbulent shear flow by modifying the definition of the constant E as given by Eqn. (2.85) (Taylor, 1954). Generally, turbulent shear flow produces a much larger velocity deviation in a cross section than that by laminar shear flow, and molecular diffusion coefficient associated with a laminar flow should be replaced by a turbulent diffusion coefficient. With these considerations, the modified Eqn. (2.85) has the following form (Fischer et al., 1979):

$$E = \frac{1}{h}\int_0^h u' \int_0^y \frac{1}{\varepsilon} \int_0^y u' dy dy dy \qquad (2.89)$$

One of the greatest contributions by G.I. Taylor (Taylor, 1954) in the field of Environmental Fluid Mechanics was his findings that the basic molecular diffusion equation can be used to simulate the substance transport in a shear flow, simply by replacing the molecular diffusion coefficient D_m with the shear-flow dispersion coefficient E.

The values of shear-flow dispersion coefficient are in general much larger than the values of molecular diffusion coefficient and turbulent diffusion coefficient (Zison et al., 1978). Therefore, only shear-flow dispersion coefficient needs to be included in Eqn. (2.88). Generally, the diffusive mass flux in rivers and groundwater aquifers, which involve shear flows, can be simulated by considering shear flow dispersion coefficient only. Under these circumstances, the shear flow dispersion coefficient is simply referred to as the dispersion coefficient.

In the simulation of the transport of dissolved and suspended substances in a two-dimensional river or groundwater, the longitudinal transport subjects to the effects of advection as well as longitudinal dispersion, and the lateral transport mainly subjects to lateral dispersion. For a conservative substance, the governing equation for a two-dimensional river transport model has the following form:

$$\frac{\partial C}{\partial t} + u \frac{\partial C}{\partial x} = E_x \frac{\partial^2 C}{\partial x^2} + E_y \frac{\partial^2 C}{\partial y^2} \qquad (2.90)$$

where E_x and E_y are longitudinal and lateral dispersion coefficients, respectively. Ideally, the values of these coefficients for river and groundwater modeling should be determined in field surveys using the tracer method (see Sections 7.2.3 and 9.2.5).

2.4   PHYSICALLY BASED WATER ENVIRONMENT MODELING

2.4.1   Simulation of mass transport mechanisms

The advection–turbulent diffusion equation or Eqn. (2.69) can be readily applied to simulation of the transport of conservative substance in turbulent water environment. Because Eqn. (2.69) is a linear PDE, the effect of reaction on substance transport can be superimposed to the equation.

Therefore, nonconservation substances in turbulent water environment can be formulated as follows:

$$\frac{\partial C}{\partial t} = -u\frac{\partial C}{\partial x} - v\frac{\partial C}{\partial y} - w\frac{\partial C}{\partial z} + \varepsilon_x\frac{\partial^2 C}{\partial x^2} + \varepsilon_y\frac{\partial^2 C}{\partial y^2} + \varepsilon_z\frac{\partial^2 C}{\partial z^2} + r \qquad (2.91)$$

Nonconservative substances may undergo various physical, chemical, and biological reactions in water environment. Generally, r in Eqn. (2.91) may consist of many terms. In Chapter 4, the mathematical formulation of a few common reaction kinetics and the determination of the values of reaction rate constants based on experimental data will be discussed.

DO in a water body is the result of advection, dispersion, and many relevant reaction kinetics. By including these reaction kinetics, a typical water environment model of DO variation in a river can be formulated as follows:

$$\frac{\partial D}{\partial t} + u\frac{\partial D}{\partial x} = E_x\frac{\partial^2 D}{\partial x^2} + E_y\frac{\partial^2 D}{\partial y^2} + \left(k_cL_c + k_nL_n - k_2D + S_b - P_o + R\right) (2.92)$$

where D represents the DO deficit (g/m^3), and $D = DO_s - DO$ with DO_s being DO saturation; k_c is the deoxygenation coefficient of carbonaceous substances; L_c is the carbonaceous biochemical oxygen demand (CBOD); k_n is the deoxygenation coefficient of nitrogenous substances; L_n is the nitrogenous biochemical oxygen demand (NBOD); k_2 is the reaeration coefficient; S_b is the benthic oxygen demand; P_o is the rate of oxygen production by aquatic plants; and R is the rate of oxygen reduction or respiration by aquatic plants. In Chapter 6, the water quality model will be used to discuss the changing mechanisms of DO and the reaction kinetics.

In Chapter 7, the determination of the values of reaction coefficients in Eqn. (2.92) based on the field data collected by an intensive water quality survey will be discussed. In a free-flowing river, advection is predominant in longitudinal direction and dispersion is in both longitudinal and lateral directions. Deoxygenation that reduces DO by carbonaceous and nitrogenous substances and reaeration that increases DO by air–water interface transfer are two primary processes. For a biologically active river, photosynthesis and respiration by aquatic plants also affect DO balance.

2.4.2 Simulation of hydrodynamic mechanisms

A hydrodynamic model is made up by conservation of mass equation or continuity equation, and conservation of momentum equations or equations of motion. As water can be treated as an incompressible fluid, continuity equation of water flow takes the following form:

$$\frac{\partial u}{\partial x} + \frac{\partial v}{\partial y} + \frac{\partial w}{\partial z} = 0 \qquad (2.93)$$

The equation of motions, also known as the Navier–Stokes equations, can be expressed as follows:

$$\frac{\partial u}{\partial t} + u\frac{\partial u}{\partial x} + v\frac{\partial u}{\partial y} + w\frac{\partial u}{\partial z} = -\frac{1}{\rho}\frac{\partial p}{\partial x} + g_x + \frac{\mu_m}{\rho}\nabla^2 u \tag{2.94}$$

$$\frac{\partial v}{\partial t} + u\frac{\partial v}{\partial x} + v\frac{\partial v}{\partial y} + w\frac{\partial v}{\partial z} = -\frac{1}{\rho}\frac{\partial p}{\partial y} + g_y + \frac{\mu_m}{\rho}\nabla^2 v \tag{2.95}$$

$$\frac{\partial w}{\partial t} + u\frac{\partial w}{\partial x} + v\frac{\partial w}{\partial y} + w\frac{\partial w}{\partial z} = -\frac{1}{\rho}\frac{\partial p}{\partial z} + g_z + \frac{\mu_m}{\rho}\nabla^2 w \tag{2.96}$$

where ρ is the fluid density; p is the water pressure; g_x, g_y, and g_z are gravitational acceleration in x, y, and z directions; and μ_m is viscosity. In these equations, fluid density and viscosity are constants. The four terms on the left-hand side of these equations denote fluid acceleration, and the three terms on the right-hand side denote external forces.

Equations (2.94)–(2.96) of motion are for laminar flow. However, water environment such as rivers, lakes, and coastal waters are generally under turbulent flow conditions. As discussed in Section 2.3.1, at any point in a turbulent flow, the instantaneous velocity is made up of average velocity and fluctuating velocity (Figure 2.6). Further understanding of fluctuating velocity has been an important subject in fluid mechanics. One major development is to introduce fluctuating velocity into Eqns. (2.94)–(2.96) and then take ensemble averages of the resulting equations. As a result, a Reynolds stress, similar to viscous stress, is obtained. Viscosity is the fluid property and can be considered as a constant for a particular fluid. However, eddy viscosity depends on water flow and is usually an unknown variable. Therefore, by considering eddy viscosity, hydrodynamic models would have more unknowns than the number of equations. This closure problem has made a full theoretical solution impossible. At this time, hydrodynamic model of turbulent flow is handled by introducing additional conceptual formulations. The $k - \varepsilon$ method, which is one of these conceptual formulations, adds into the classical hydrodynamic model a conservation of energy equation, which includes turbulence kinetic energy k and turbulence energy dissipation rate ε (Clark, 1996).

2.4.3 Mathematical formulation of physically based water environment models

Flow velocity and dispersion vary with hydrodynamic properties of water environment. In a complete water environment modeling, it is necessary to conduct hydrodynamic simulation first. The calculated flow velocity and dispersion coefficients are then introduced into transport equation to simulate environmental quality. Therefore, water environment modeling is often conducted in two steps sequentially (Figure 2.10).

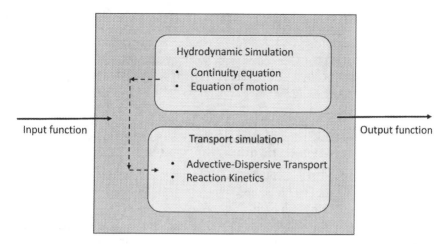

Figure 2.10 Physically based water environment modeling.

Salinity in coastal groundwater changes due to flow and dispersion, which can be calculated by solving the relevant salinity transport model. Conversely, flow velocity and dispersion coefficient in coastal water may change due to significant change of its salinity. Therefore, the groundwater flow and salinity transport in a coastal aquifer are interrelated. Under this situation, water environment modeling must be conducted by coupling transport and hydrodynamic models, which is much more complicated (Gingerich and Voss, 2005).

2.5 SYSTEM-BASED WATER ENVIRONMENT MODELING

2.5.1 Mathematical formulation of system-based models

The concept of systems modeling was first proposed by Volterra (1959) for the simulation of electromagnetic and elastic processes. Volterra showed that, as an important common feature, the present states of electromagnetic and elastic processes are related to their past experience as well as basic system properties. He further showed that this relationship can be expressed by an infinite integral series in the form of Eqn. (2.97). Since then, this infinite integral series has been used widely in various engineering analyses and is called the Volterra integral equation.

$$
\begin{aligned}
y(t) = y_0(t) &+ \int_{-\infty}^{t} h_1(t,s)x(s)ds \\
&+ \int_{-\infty}^{t}\int_{-\infty}^{t} h_2(t,s_1,s_2)\, x(s_1)x(s_2)ds_1ds_2 + \cdots \\
&+ \int_{-\infty}^{t}\cdots\int_{-\infty}^{t} h_2(t,s_1,s_2\cdots,s_n)\prod_{n}^{i=1}x(s_i)ds_i
\end{aligned}
\tag{2.97}
$$

where $y(\ )$ and $x(\ )$ are system output and input functions, respectively. The first term on the left-hand side or $y_0(t)$ represents the effects of initial state on the output function, and $h_1, h_2, \ldots h_n$ are the kernel functions that represent the basic properties of the system. In engineering hydrology, Volterra integral equation was used successfully to formulate nonlinear watershed rainfall–runoff models (Liu and Brutsaert, 1978; Xia, 2002).

2.5.2 Compatibility of physically based and system-based modeling approaches

In Section 2.4, the mathematical formulation of water environment models by following a physically based approach is discussed. These models consist of governing equations and boundary/initial conditions. In general, governing equations are in the form of differential equations that simulate transporting mechanisms of advection, diffusion or dispersion, and reaction.

An ordinary differential equation can be put in a general form as follows:

$$\Gamma\phi(x) = f(x) \tag{2.98}$$

where Γ is an ordinary differentiation operator, $\phi(x)$ is an unknown variable, and $f(x)$ is a known variable. If an inverse differentiation operator Γ^{-1} exists such that $\Gamma\Gamma^{-1} \equiv 1$, the solution of Eqn. (2.98) can be readily obtained as follows:

$$\phi(x) = \Gamma^{-1}f(x) \tag{2.99}$$

Clearly, the success of the aforementioned solution depends on the determination of the inverse differentiation operator Γ^{-1}. An inverse operation of differentiation is integration. By defining $G(\omega)$ as a kernel of integration function, the corresponding integration can be expressed as follows:

$$\Gamma^{-1}f(x) = \int_{\Omega} G(\omega)f(x - \omega)d\omega \tag{2.100}$$

where Ω is the region of integration, ω is integration variable, G is also called the Green's function of a differentiation operation.

Green's function is an important concept in mathematical physics. Using the concept of Green's function, as shown earlier, a boundary value problem of differentiation equation can be transformed into a problem of integral equation.

With Eqn. (2.100), the solution to the original differential equation or Eqn. (2.98) becomes:

$$\phi(x) = \Gamma^{-1}f(x) = \int_{\Omega} G(\omega)f(x - \omega)d\omega \tag{2.101}$$

Changing the integration coefficient, Eqn. (2.101) can be expressed as follows:

$$\phi(x) = \Gamma^{-1}f(x) = \int_{\Omega} G(x-\omega)f(\omega)d\omega \tag{2.102}$$

Further manipulation leads to:

$$\begin{aligned}\phi(x) &= I\phi(x) = \Gamma\Gamma^{-1}\phi(x) = \Gamma\int_{\Omega} G(x-\omega)\phi(\omega)d\omega \\ &= \int_{\Omega} \Gamma G(x-\omega)\phi(\omega)d\omega\end{aligned} \tag{2.103}$$

where $I = \Gamma\Gamma^{-1}$ is called an identity operator.

The definition and application of a Dirac delta function $d(x)$ has been introduced and discussed earlier in Section 2.2.1. With a Dirac delta function, any continuous function $\xi(x)$ can be expressed as follows (Sokolnikoff and Redheffer, 1966):

$$\xi(x) = \int_{\Omega} \delta(x-\omega)\xi(\omega)d\omega \tag{2.104}$$

As $\xi(x)$ is a continuous function, by comparing Eqns. (2.101) and (2.104), the following useful conclusion is reached: when the known function $f(x)$ of a given differential equation is a Dirac delta function, Green's function is the solution of the differential equation.

By comparing Eqns. (2.103) and (2.104), we have:

$$\Gamma G(x-\omega) = \delta(x-\omega) \tag{2.105}$$

which can be simplified to:

$$\Gamma G(x) = \delta(x) \tag{2.106}$$

or

$$G(x) = \Gamma^{-1}\delta(x) \tag{2.107}$$

After the Green's function of a differential equation is obtained, the solution to the differential equation with known function of $f(x)$ can be readily calculated by inserting the Green's function into Eqn. (2.105). For simplicity, the earlier derivation was conducted with respect to an ordinary differential equation. These derivations can be readily extended to the PDE (Friedman, 1990).

Green's function is the solution of a differential equation when the known function or the input function is a Dirac delta function. Furthermore, a water environment model in the form of differential equations can be transformed into a model in the form of an integral equation. The model in the form of differential equation is a physically based model as it simulates explicitly the actions of hydrodynamic and transport mechanisms. The model in terms of integral equation is a system-based model as it simulates the action of hydrodynamic and transport mechanisms implicitly in terms of system's response function.

2.5.3   System identification in linear system water environment modeling

Linear system model can be formulated by keeping only the first two terms of a Volterra integral series (Liu and Brutsaert, 1975):

$$y(t) = x_0(t) + \int_0^t h(t,s)x(s)ds \tag{2.108}$$

As shown in Figure 2.11, the output function of a linear system model consists of two independent parts: (1) zero-input response that is produced by the initial state of the system and (2) zero-state response that is produced by the system input or loadings.

A linear system is a time-invariant system if it produces the same output after receiving the same input, regardless of when the input occurs. For a time-invariant system, the kernel function in Eqn. (2.108), $h(t,s)$ is only a function of $(t-s)$ and can be denoted as $h(t-s)$ (Volterra, 1959). Also, in linear system modeling, the effects of the initial state on the output are often small and can be ignored. Under these two conditions, Eqn. (2.108) becomes:

$$y(t) = \int_0^t h(t-s)x(s)ds \tag{2.109}$$

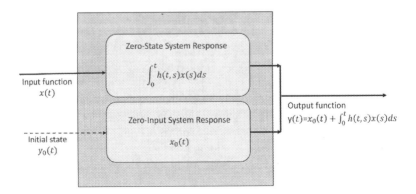

Figure 2.11 Linear system water environment modeling.

Equations (2.109) and (2.102) are similar, and thus, the kernel function of a linear systems model is the Green's function of a corresponding physically based model. When Eqn. (2.109) is used for the simulation of a watershed rainfall–runoff system, the kernel function is referred to as an impulse response function or an instantaneous unit hydrograph (IUH; Liu and Brutsaert, 1975).

The process of evaluating the impulse response function of a linear system is called system identification. Successful linear system modeling largely depends on how efficient and accurate the system identification can be. System identification of a water environment system can be accomplished by using one of the following three methods: the method of physical parameterization, the inverse method, and the method of system parameterization (Liu, 1988).

2.5.3.1 *The method of physical parameterization*

Based on the method of physical parameterization, the impulse response function of the linear systems model of a water environment system is expressed in terms of coefficients that denote transport and hydrodynamic mechanisms. In its application, a physically based model is first formulated, and the impulse response function can be derived as the solution of this model with a Dirac delta function input. Therefore, the application of this method requires prior knowledge of the transport and hydrodynamic mechanisms of a water environment system. This method is also called a white-box method of system identification.

As shown in Section 2.2.1, Eqn. (2.24) is the analytical solution of a basic molecular diffusion model with the Dirac delta function as the model's initial conditions. Therefore, Eqn. (2.24) is the impulse response of the basic molecular diffusion model.

Example 2.5 System identification by the method of physical parameterization.

Problem: A water body can be simulated by a basic diffusion model. (1) Formulate a linear systems model of the water body and (2) determine the impulse response function of the model (system identification) by the method of system parameterization.

Solution

1. The general form of a linear system model of a water environment is as follows:

$$y(t) = \int_{0}^{t} h(t-s)x(s)\,ds$$

2. By the method of system parameterization, the system response function of a linear system model is the solution to the physically based

model of this water body. Section 2.2.1 shows that the basic molecular diffusion model consists of Eqn. (2.15) as its governing equation:

$$\frac{\partial C}{\partial t} = D_m \frac{\partial^2 C}{\partial x^2}$$

and the Dirac delta function Eqn. (2.16) as the boundary/initial conditions:

$$C(x,0) = M\delta(0)$$

Section 2.2.1 also shows that the solution to the basic molecular diffusion model is Eqn. (2.24):

$$C(x,t) = \frac{M}{\sqrt{4\pi D_m t}} e^{-\left(\frac{x^2}{4D_m t}\right)}$$

The system response function of a linear system model $h(t)$ for this water body is Eqn. (2.24) with the mass input being equal to unity or $M = 1$. Therefore,

$$h(x,t) = \frac{1}{\sqrt{4\pi D_m t}} e^{-\left(\frac{x^2}{4D_m t}\right)}$$

Example 2.6 Application of a linear system model.
Problem: Solve the linear system model of Example 2.5, if this water body receives initially (at $t = 0$) an input of a waste substance in a concentration distribution $C(x,0) = f(x)$, as shown in Figure 2.12.
Solution: Because the model's governing equation or Eqn. (2.15) is a linear equation, the model solutions with other boundary/initial conditions can be obtained readily by using the superposition principle.

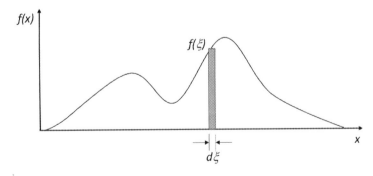

Figure 2.12 Application of the linear system modeling.

By considering a small element of the input function, as shown by the shaded area in Figure 2.12, the mass of this element area, which is located at a distance $x = \xi$, is $f(\xi)d\xi = M$ where $d\xi$ is very small. Thus, the boundary/initial conditions can be expressed approximately by a Dirac delta function in the following form:

$$C(x,0) = M\delta(x - \xi) \tag{2.110}$$

The solution to a molecular diffusion model with the boundary/initial conditions of Eqn. (2.110) can be expressed as follows:

$$C(x,t) = -\frac{M}{\sqrt{4\pi D_m t}} e^{\left[\frac{-(x-\xi)^2}{4D_m t}\right]} \tag{2.111}$$

The input function can be considered as a series of small element $f(\xi)d\xi$ based on the linear system theory. The tracer substance distribution in a one-dimensional molecular diffusion system is the following convolution integration:

$$C(x,t) = -\int_{-\infty}^{\infty} \frac{f(\xi)}{\sqrt{4\pi D_m t}} e^{\left[\frac{-(x-\xi)^2}{4D_m t}\right]} d\xi \tag{2.112}$$

Comments

1. A water environment system can be simulated by following a physically based approach or alternatively by a systems approach. The solution of a physically based model receiving a pulse input is essentially the system's impulse response function. Thus, the use of the method of system parameterization in system modeling approach is similar to the use of physically based modeling approach. However, as discussed in Section 8.4.3, the conjunctive use of two modeling approaches may yield better understanding of the hydrodynamic and reaction mechanisms of a water environment system.
2. The temporal and spatial distribution of a waste substance in a linear transport system, after it receives a waste input of arbitrary spatial distribution $f(x)$ at $t = 0$, can be expressed by Eqn. (2.112).

2.5.3.2 The inverse method

If both input and output functions of a given linear transport system for a historical event are available, the system impulse response can be calculated by the inverse convolution integration. Therefore, application of this method requires no prior knowledge of the transport and hydrodynamic mechanisms of a water environment system. This method is also called a black-box method of system identification. This method has been widely used in watershed hydrology for the determination of IUH with given rain-fall (input) and direct runoff (output; Liu and Brutsaert, 1975). Detail discussion is given in Chapter 5.

2.5.3.3 The method of system parametrization

Collection of rainfall and runoff data is often subjected to uncertainties. Therefore, the impulse response function of IUH of a watershed using the inverse method subjects to the weakness of robustness – the calculated IUH reproduces the historical hydrologic event, but it often calculates the future hydrologic event less than satisfaction. One way to resolve this weakness is to assume the IUH of a watershed to be of a particular statistical distribution, which is characterized by several parameters. Instead of calculating an IUH of a watershed directly based entirely on historical rainfall and runoff data, these historical data are used to determine the values of the parameters of assumed statistical distribution. Therefore, application of this method assumes partial knowledge of the transport and hydrodynamic mechanisms of a water environment system. This method is also called a grey-box method of system identification. In linear system modeling of a watershed system, Nash (1959) assumed that the IUH took a general form of two-parameter Gamma distribution (Chapter 5). In linear system modeling of solute transport in soils, Jury (1982) assumed that the impulse response function, which he called as the transfer function, took the form of a two-parameter log-normal distribution (Chapter 8).

EXERCISES

1. Ethanol is released initially ($t = 0$) into a still water and forms a thin layer with a concentration of 10 g/cm^2. Molecular diffusion is the only mixing mechanism in this water body, and the molecular diffusion coefficient is $D_m = 1.0$ cm^2/s.

    (a) Calculate and plot the ethanol concentration distribution at $t = 1.0$ s, $t = 5.0$ s, and $t = 2.0$ s.
    (b) Plot the variation of maximum ethanol concentration with time, or C_{max} versus t.

2. For the same still water body as the last problem, plot the variation of maximum ethanol concentration with time, or C_{max} versus t, if the initial ethanol layer is 3.0 cm thick.

3. A field experiment is conducted in a stream to measure its dispersion coefficient, as shown in Figure 2.13.

    The tracer is released initially ($t = 0$) into the stream as an instantaneous line source. The transport of the tracer in the stream can be

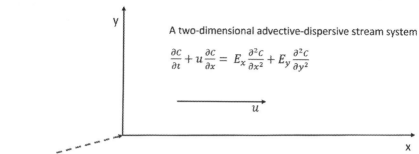

y

A two-dimensional advective-dispersive stream system

$$\frac{\partial C}{\partial t} + u\frac{\partial C}{\partial x} = E_x\frac{\partial^2 C}{\partial x^2} + E_y\frac{\partial^2 C}{\partial y^2}$$

u

x

Tracer released as an instantaneous line source

Figure 2.13 Field measurement of river dispersion coefficient (Exercise problem 3).

simulated by a two-dimensional advection–dispersion model. The model governing equation and boundary/initial conditions can be expressed as follows:

Governing equation: $\dfrac{\partial C}{\partial t} + u\dfrac{\partial C}{\partial x} = E_x\dfrac{\partial^2 C}{\partial x^2} + E_y\dfrac{\partial^2 C}{\partial y^2}$

Initial/boundary conditions: $C(x,0) = M\delta(0)$

Verify that the solution of this tracer transport model is obtained as follows:

$$C(x,y,t) = \frac{M}{4\pi t\sqrt{E_x E_y}}e^{\left[\frac{(x-ut)^2}{4E_x t}\frac{y^2}{4E_y t}\right]}$$

4. A glass cup is filled with pure water of 10 cm in depth. When an experiment starts, a very thin layer of salty water is added carefully at the cup bottom so that no turbulence was generated, and molecular diffusion is the only mixing mechanism. Assume molecular coefficient is 0.1 cm²/s (as shown in Figure 2.14). Calculate how long it takes for the salinity at the water surface to become 25% of that at the bottom.

   Hint: Solve the problem by using the method of images as introduced in Example 2.4.

5. A still river is separated from a still lake by a sluice gate. River water contains a contaminant of concentration C_0, while the lake water is completely free of this contaminant. The gate is opened at $t = 0$ and the contaminant transport from the river to lake by molecular diffusion. To apply the linear system modeling technique (refer to Example 2.5) and

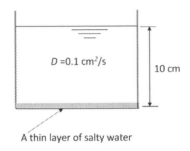

Figure 2.14 Application of the method of imaging (Exercise problem 4).

verify that the contaminant distribution in the lake can be expressed by the following equation:

$$C(x,t) = C_0 \left[1 - \mathrm{erf}\left(\frac{x}{\sqrt{4D_m t}} \right) \right]$$

where x is the distance from the sluice gate, which separates the river.

REFERENCES

Clark, M.M. (1996). *Transport Modeling for Environmental Engineers and Scientists.* John-Wiley & Sons, New York.

Crank, J. (1975). *The Mathematics of Diffusion.* Oxford University Press, London and New York.

Csanady, G.T. (1973). *Turbulent Diffusion in the Environment.* D. Reidel Publishing Company, Dordrecht-Holland and Boston, MA.

Daily, J.C. and Harleman, D.F. (1972). *Fluid Dynamics.* Addison-Wesley Publishing Company, Reading, MA.

Fischer, H.B., List, E., Koh, R., Imberger, J. and Brooks, N. (1979). *Mixing in Inland and Coastal Waters.* Academic Press, New York, pp. 229–242.

Friedman, B. (1990). *Principles and Techniques of Applied Mathematics.* Dover, New York.

Gingerich, S.B. and Voss, C.I. (2005). Three-dimensional variable-density flow simulation of a coastal aquifer in Southern Oahu, Hawaii, USA. *Hydrogeology Journal*, 13(2), pp. 436–450.

Jury, W.A. (1982). Simulation of solute transport using a transfer function model. *Water Resources Research*, 18(2), pp. 363–368.

Kotz, J.C., Treichel, P. and Townsend, J.R. (2009). Bonding and molecular structure. Part 2 in *Chemistry and Chemical Reactivity*, 7th Edition. Brooks/Cole Publishing, Boston, MA.

Liu, C.C.K. (1988). Solute transport modeling in heterogeneous soils: Conjunctive application of physically based and the system approaches. *Journal of Contaminant Hydrology*, 3(1), pp. 97–111.

Liu, C.C.K. and Brutsaert, W.H. (1975). Optimal identification of a watershed rain-runoff system. *IAHS-AISH Publication No. 117*, International Association of Hydrological Sciences, Wallingford, UK, pp. 333–341.

Liu, C.C.K. and Brutsaert, W.H. (1978). A nonlinear analysis of the relationship between rainfall and runoff for extreme floods. *Journal of Water Resources Research*, 14(1), pp. 75–83.

Massey, B.S. (1983). *Mechanics of Fluids*, 5th Edition. English Language Book Society and Van Nortrand Reinhold Co., United Kingdom.

Nash, J.E. (1959). Systematic determination of unit hydrograph parameters. *Journal of Geophysical Research*, 64(1), pp. 111–115.

Rohsenow, W.M. and Choi, H.Y. (1961). *Mass and Momentum Transfer*. Prentice Hall Inc, New York.

Sokolnikoff, I.S. and Redheffer, R.M. (1966). *Mathematics of Physics and Modern Engineering*. McGraw-Book Book Co, New York.

Taylor, G.I. (1921). Experiments with rotating fluids. *Proceedings of the Royal Society of London. Series A, Containing Papers of a Mathematical and Physical Character*, 100(703), pp. 114–121.

Taylor, G.I. (1954). The dispersion of matter in turbulent flow through a pipe. *Proceedings of the Royal Society of London. Series A. Mathematical and Physical Sciences*, 223(1155), pp. 446–468.

Volterra, V. (1959). *Theory of Functionals and of Integral and Integro-differential Equations*. Dover, New York.

Xia, J. (2002). A system approach to real-time hydrologic forecast in watersheds. *Water International*, 27(1), pp. 87–97.

Zison, S.W., Mills, W.B., Deimer, D. and Chen, C.W. (1978). *Rates, Constants, and Kinetics Formulation in Surface Water Quality Modeling*. U.S. Environmental Protection Agency Research Report EPA-600/3-78-105.

Chapter 3

Numerical methods for water environment modeling

3.1 ANALYTICAL AND NUMERICAL SOLUTIONS OF WATER ENVIRONMENT MODELS

In a water environment model, the hydrodynamic parameters such as flow velocity and diffusion coefficient are often taken as input constants to simplify the problem setup. In such a case, the water environment model contains only the mass transport equation. When the mass transport equation is solved, its analytical solutions can be obtained by the method introduced in Sections 2.2 and 2.3 under special boundary and initial conditions.

From 1932, the Streeter–Phelps model has been widely used to calculate the river self-purification capacity for organic pollutants and the allowable discharge of pollutants from the sewage treatment plant. The Streeter–Phelps model has been used for rivers under low flow periods when the flows are steady. The flow velocity can be assumed to be constant, and the dispersion process can be neglected so that the analytical solution can be obtained. Although the mode is simple, it leads to a deeper understanding of the hydrodynamic and natural response mechanisms of the water environment. Therefore, such simple model should be derived and applied first whenever it is possible.

However, for more complex problems such as unsteady inflow or irregular river channels, the mass transport equation and hydrodynamic equation in a water environment model require numerical solution in general. When the equations are solved numerically, the computational domain is discretized into small cells where the partial differential governing equations are approximated by a set of algebraic equations that are solved by iteration or the matrix inversion method. The most commonly used numerical methods in water environment modeling analysis are the FDM, the FEM, the method of characteristics and the boundary element method. In this chapter, we shall introduce the FDM and the FEM first, followed by two popular water environment modeling tools of Hydrologic Engineering Center-River Analysis System (HEC-RAS; for river flow) and SUTRA (for groundwater flow), which use the FDM and the FEM as numerical techniques, respectively.

DOI: 10.1201/9781003008491-3

3.2 FINITE DIFFERENCE METHOD

The FDM is a numerical method to solve a PDE by using finite difference (FD) schemes to approximate local derivatives and to seek the approximate solution to the original PDE.

3.2.1 Taylor expansion and scheme construction

Let's first use the function $\phi(x) = x^2$ as the example to illustrate how an FD scheme is constructed for $\partial\phi/\partial x$ and $\partial^2\phi/\partial x^2$. The notations ϕ_{i-1}, ϕ_i, and ϕ_{i+1} represent the value of ϕ at the nodes $i-1$, i, and $i+1$, respectively. The distance between node $i-1$ and i is defined as $\Delta x_{i-1/2}$ and between i and $i+1$ as $\Delta x_{i+1/2}$, as shown in Figure 3.1.

The common approach to develop an FD scheme starts from Taylor expansion, which gives:

$$
\begin{aligned}
\phi_{i-1} = \phi_i &+ \left(-\Delta x_{i-1/2}\right)\left(\frac{\partial\phi}{\partial x}\right)_i + \frac{\left(-\Delta x_{i-1/2}\right)^2}{2!}\left(\frac{\partial^2\phi}{\partial x^2}\right)_i \\
&+ \frac{\left(-\Delta x_{i-1/2}\right)^3}{3!}\left(\frac{\partial^3\phi}{\partial x^3}\right)_i + \frac{\left(-\Delta x_{i-1/2}\right)^4}{4!}\left(\frac{\partial^4\phi}{\partial x^4}\right)_i \\
&+ \cdots + \frac{\left(-\Delta x_{i-1/2}\right)^m}{m!}\left(\frac{\partial^m\phi}{\partial x^m}\right)_i + \cdots
\end{aligned}
\tag{3.1}
$$

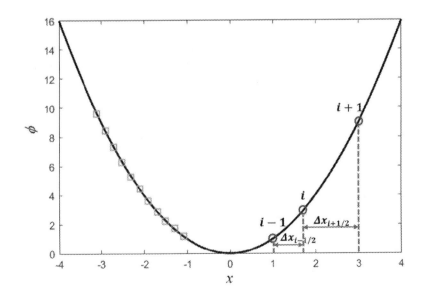

Figure 3.1 FDM representation of a continuous function by discrete points.

$$\phi_{i+1} = \phi_i + \left(\Delta x_{i+1/2}\right)\left(\frac{\partial \phi}{\partial x}\right)_i + \frac{\left(\Delta x_{i+1/2}\right)^2}{2!}\left(\frac{\partial^2 \phi}{\partial x^2}\right)_i$$

$$+ \frac{\left(\Delta x_{i+1/2}\right)^3}{3!}\left(\frac{\partial^3 \phi}{\partial x^3}\right)_i + \frac{\left(\Delta x_{i+1/2}\right)^4}{4!}\left(\frac{\partial^4 \phi}{\partial x^4}\right)_i$$

$$+ \cdots + \frac{\left(\Delta x_{i+1/2}\right)^m}{m!}\left(\frac{\partial^m \phi}{\partial x^m}\right)_i + \cdots \qquad (3.2)$$

The above Taylor expansion is valid if the function and its derivatives are smooth.

Now a particular FD scheme can be derived from the above Taylor expansion. First, let's use two points ϕ_{i-1} and ϕ_i to approximate $(\partial \phi / \partial x)_i$ with their linear combination:

$$\left(\partial \phi / \partial x\right)_i = a\phi_{i-1} + b\phi_i$$

$$= a\left[\begin{array}{l}\phi_i + \left(-\Delta x_{i-1/2}\right)\left(\frac{\partial \phi}{\partial x}\right)_i + \frac{\left(-\Delta x_{i-1/2}\right)^2}{2!}\left(\frac{\partial^2 \phi}{\partial x^2}\right)_i \\ + \cdots + \frac{\left(-\Delta x_{i-1/2}\right)^m}{m!}\left(\frac{\partial^m \phi}{\partial x^m}\right)_i\end{array}\right] + b\phi_i$$

$$= (a+b)\phi_i + \left(-a\Delta x_{i-1/2}\right)\left(\frac{\partial \phi}{\partial x}\right)_i + \frac{a\left(-\Delta x_{i-1/2}\right)^2}{2!}\left(\frac{\partial^2 \phi}{\partial x^2}\right)_i$$

$$+ \cdots + \frac{a\left(-\Delta x_{i-1/2}\right)^m}{m!}\left(\frac{\partial^m \phi}{\partial x^m}\right)_i \qquad (3.3)$$

In order to find the values of a and b of the equation, we need to have the coefficients in front of each term being equal to each other on both sides, which gives a system of equations as follows:

$$\begin{cases} a+b=0 \\[2mm] -a\Delta x_{i-1/2} = 1 \\[2mm] \dfrac{a\left(-\Delta x_{i-1/2}\right)^2}{2!} = 0 \\[2mm] \cdots \\[2mm] \dfrac{a\left(-\Delta x_{i-1/2}\right)^m}{m!} = 0 \end{cases} \qquad (3.4)$$

The aforementioned system of equations is an overdetermined matrix with more constraints than unknowns, and thus, there is no solution in general. But if we leave out the higher derivative terms by retaining the first two equations only, the coefficients a and b can be solved, which gives:

$$a = -\frac{1}{\Delta x_{i-1/2}} \quad b = \frac{1}{\Delta x_{i-1/2}} \tag{3.5}$$

This gives the final form of the FD scheme to be:

$$(\partial \phi / \partial x)_i = \frac{\phi_i - \phi_{i-1}}{\Delta x_{i-1/2}} + O\left(\Delta x_{i-1/2}\right) \tag{3.6}$$

The aforementioned scheme is based on the local node i and its backward node $i-1$, and the scheme is called *backward difference*. By using the same method, we can use the local node i and its forward node $i+1$ to construct the *forward difference*:

$$\left(\frac{\partial \phi}{\partial x}\right)_i = \frac{\phi_{i+1} - \phi_i}{\Delta x_{i+1/2}} + TE \tag{3.7}$$

where TE is truncation errors, which will be detailed later. We can also use its forward node $i+1$ and its backward node $i-1$ to construct central difference, where the derivative is evaluated at the location between two nodal points.

$$\left(\frac{\partial \phi}{\partial x}\right)_i = \frac{\phi_{i+1} - \phi_{i-1}}{\Delta x_{i-1/2} + \Delta x_{i+1/2}} + TE \tag{3.8}$$

The use of Taylor expansion not only provides the theoretical explanation for the error order but also serves as the basis for the systematical development of other more complex FD schemes.

3.2.2 Truncation error and order of accuracy

In practical computation, the characteristics of the errors for different FD scheme are different, which can be obtained by analyzing the characteristics of the truncated higher-order term (truncation error). The TE is defined as the difference between the derivative to be solved and its FD expression.

Taking the backward FD scheme, for example, the truncation error is expressed as follows:

$$
\begin{aligned}
TE &= \left(\frac{\partial \phi}{\partial x}\right)_i - \frac{\phi_i - \phi_{i-1}}{\Delta x_{i-1/2}} \\
&= \left(\frac{\partial \phi}{\partial x}\right)_i - \frac{1}{\Delta x_{i-1/2}} \left\{ \phi_i - \left[\begin{array}{l} \phi_i + (-\Delta x_{i-1/2})\left(\frac{\partial \phi}{\partial x}\right)_i + \frac{(-\Delta x_{i-1/2})^2}{2!}\left(\frac{\partial^2 \phi}{\partial x^2}\right)_i \\ + \frac{(-\Delta x_{i-1/2})^3}{3!}\left(\frac{\partial^3 \phi}{\partial x^3}\right)_i \\ + \frac{(-\Delta x_{i-1/2})^4}{4!}\left(\frac{\partial^4 \phi}{\partial x^4}\right)_i \\ + \cdots + \frac{(-\Delta x_{i-1/2})^m}{m!}\left(\frac{\partial^m \phi}{\partial x^m}\right)_i + \cdots \end{array} \right] \right\} \\
&= \frac{\Delta x_{i-1/2}}{2!}\left(\frac{\partial^2 \phi}{\partial x^2}\right)_i - \frac{(\Delta x_{i-1/2})^2}{3!}\left(\frac{\partial^3 \phi}{\partial x^3}\right)_i + \frac{(\Delta x_{i-1/2})^3}{4!}\left(\frac{\partial^4 \phi}{\partial x^4}\right)_i + \cdots
\end{aligned} \tag{3.9}
$$

The truncation error term reflects the error caused by using discrete points to calculate the derivative. It not only gives the accuracy of the FD scheme, which is the lowest order in the error terms, but also gives the specific expression of each error term.

By using the same method, the TE for forward difference scheme can be obtained:

$$
\begin{aligned}
TE &= \left(\frac{\partial \phi}{\partial x}\right)_i - \frac{\phi_{i+1} - \phi_i}{\Delta x_{i+1/2}} \\
&= -\frac{\Delta x_{i+1/2}}{2!}\left(\frac{\partial^2 \phi}{\partial x^2}\right)_i - \frac{(\Delta x_{i+1/2})^2}{3!}\left(\frac{\partial^3 \phi}{\partial x^3}\right)_i - \frac{(\Delta x_{i+1/2})^3}{4!}\left(\frac{\partial^4 \phi}{\partial x^4}\right)_i - \cdots
\end{aligned} \tag{3.10}
$$

Although the order of accuracy is the same for both forward and backward difference scheme, the sign in front of all even order derivatives (diffusion) are different, which changes the characteristics of two FD schemes.

The TE for central difference is expressed as follows:

$$
\begin{aligned}
TE &= \left(\frac{\partial \phi}{\partial x}\right)_i - \frac{\phi_{i+1} - \phi_{i-1}}{\Delta x_{i-1/2} + \Delta x_{i+1/2}} \\
&= -\frac{\Delta x_{i+1/2} - \Delta x_{i-1/2}}{2!}\left(\frac{\partial^2 \phi}{\partial x^2}\right)_i - \left[\frac{(\Delta x_{i+1/2})^3 + (\Delta x_{i-1/2})^3}{3!(\Delta x_{i-1/2} + \Delta x_{i+1/2})}\right]\left(\frac{\partial^3 \phi}{\partial x^3}\right)_i \\
&\quad - \left[\frac{(\Delta x_{i+1/2})^4 - (\Delta x_{i-1/2})^4}{4!(\Delta x_{i-1/2} + \Delta x_{i+1/2})}\right]\left(\frac{\partial^4 \phi}{\partial x^4}\right)_i - \cdots
\end{aligned} \tag{3.11}
$$

It can be observed that when the nonuniform grid is used (i.e., $\Delta x_{i+1/2} \neq \Delta x_{i-1/2}$), the scheme remains to be first-order accurate, while when uniform grids are applied (i.e., $\Delta x_{i+1/2} = \Delta x_{i-1/2}$), the central difference scheme becomes second-order accurate. In order to have a second-order FD for $(\partial \phi / \partial x)_i$ in a nonuniform grid system, at least three grid points are needed.

3.2.3 Consistency and convergence

Consistency: Based on the introduction in the last section, any spatial derivative term can be approximated by an FD equation at a certain order of accuracy with the error terms being represented by TE. By using the same method, the temporal derivative term can also be approximated. The consistency refers to the fact that an FD scheme gives approximation close enough to a PDE so that the total TE approaches to zero when x and t approach to zero. Generally, consistency is the minimum requirement for a valid numerical scheme. If the FD for each derivative term in a PDE has at least first-order accuracy, when x and t approach to zero, the FD will be consistent with the PDE.

Convergence: The convergence shows that the numerical solution for an FD scheme is close enough to the true solution of the original PDE. Generally, convergence is a higher requirement than consistency for an FD scheme. When a scheme is convergent, it is always consistent. However, due to the in-built instability, a consistent scheme may not be convergent under certain conditions. In most of practical applications, the convergence and consistency for an FD scheme has the same meaning. As the refinement of grid size and time step, the numerical solution of an FD scheme can converge to the true solution of a PDE within a bounded finite truncation error.

3.2.4 Stability

Stability analysis is the most important step in solving an unsteady problem using an FD method, because even if the order of accuracy for an unstable scheme is high, it may lose its stability during the calculation, leading to exponential amplification of any error perturbation.

3.2.4.1 Heuristic analysis of numerical stability

Now the 1D advection equation is used as an example to illustrate how to establish heuristic error analysis for an FD scheme:

$$\frac{\partial C}{\partial t} + u \frac{\partial C}{\partial x} = 0 \tag{3.12}$$

The classical example is the 'forward time central space (FTCS)' scheme:

$$\frac{C_i^{n+1} - C_i^n}{\Delta t} + u \frac{C_{i+1}^n - C_{i-1}^n}{2\Delta x} = 0 \Rightarrow C_i^{n+1} = C_i^n - \frac{u\Delta t}{2\Delta x}\left(C_{i+1}^n - C_{i-1}^n\right) \tag{3.13}$$

The aforementioned FD scheme is explicit (i.e., only the information of the previous time step is needed to calculate the value of each node at the next time step and no iteration is required) and second-order accuracy in space. To demonstrate the characteristics of this scheme, we use this scheme to calculate the following initial-value problem:

$$C = \begin{cases} 1.5 & x \le 10 \\ 0.5 & x > 10 \end{cases} \text{ when } t = 0 \tag{3.14}$$

The following parameter values are adopted, $u = 0.5$, $\Delta x = 0.3$, and $\Delta t = 0.2$. Obviously, when the speed C is constant, the analytic solution is a distribution curve of uniform motion, that is, the initial distribution of the function will shift to right without changing its shape. However, as shown in Figure 3.2, the numerical results by the aforementioned FD scheme oscillates rapidly and the fluctuation amplitude keeps increasing, which is unacceptable. This situation is called instability of the numerical scheme.

What causes the oscillation? Based on the earlier introduction, an FD is an approximation scheme to the original PDE. The difference between the true solution and the numerical result is actually the TE, which varies for

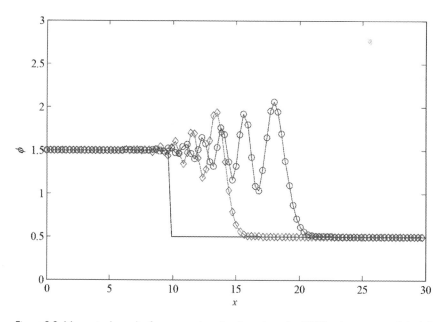

Figure 3.2 Numerical results for pure advection based on the FTCS scheme at $t = 0$ (solid line), 10 (diamond line), and 20 (circle line).

different FD schemes. For the FTCS scheme, the following analysis can be obtained:

$$
\frac{C_i^{n+1} - C_i^n}{\Delta t} + u\frac{C_{i+1}^n - C_{i-1}^n}{2\Delta x} = 0
$$
$$
\overset{equivalent}{\leftrightarrow} \left(\frac{\partial C}{\partial t}\right)_i^n + u\left(\frac{\partial C}{\partial x}\right)_i^n = TE = -\frac{\Delta t u^2}{2}\left(\frac{\partial^2 C}{\partial x^2}\right)_i^n
$$
$$
-\frac{\Delta x^2 u}{3}\left(\frac{\partial^3 C}{\partial x^3}\right)_i^n + O(\Delta x^3)
\tag{3.15}
$$

In the aforementioned analysis, the truncation error of time and space terms is unified by the approximation of $\left(\partial^2 C / \partial t^2\right)_i^n \approx u^2 \left(\partial^2 C / \partial x^2\right)_i^n$. Therefore, the TE for FTCS scheme has additional higher-order derivative terms, corresponding to the diffusion (spatial second derivative), dispersion (spatial third derivative), and other higher-order derivative terms. We know that the diffusion term always smooths the local maxima when its coefficient is positive. However, in Eqn. (3.15), the coefficient corresponding to the diffusion term is always negative, which enhances the local maxima generated by the TE and the round-off error. In this case, the numerical scheme is unstable, that is, a small error can be unlimitedly amplified as the scheme marches in time.

3.2.4.2 Von Neumann analysis of numerical stability

Another stability analysis method is called the von Neumann method, which is based on Fourier series expansion and is only applicable to linear FD schemes. At time step n for node i, an error that takes the form of one component of the Fourier series is introduced, that is:

$$
E_i^n = A^n e^{Ik_x(i\Delta x)}
\tag{3.16}
$$

where $I = \sqrt{-1}$ is imaginary unit, k_x is the wave number in the x-direction, and A^n is the corresponding amplitude.

To determine whether an FD scheme is stable, we can substitute the error term into the FD scheme and check whether the amplification factor increases at the next time step. To illustrate how this method works, we will use the FTCS scheme again as an example:

$$
\frac{C_i^{n+1} - C_i^n}{\Delta t} + u\frac{C_{i+1}^n - C_{i-1}^n}{2\Delta x} = 0 \Rightarrow C_i^{n+1} = C_i^n - \frac{u\Delta t}{2\Delta x}\left(C_{i+1}^n - C_{i-1}^n\right)
\tag{3.17}
$$

Replacing ϕ by E in the FD scheme, we obtain:

$$
\begin{aligned}
A^{n+1}e^{Ik_x(i\Delta x)} &= A^n e^{Ik_x(i\Delta x)} - \frac{u\Delta t}{2\Delta x}\left\{A^n e^{Ik_x[(i+1)\Delta x]} - A^n e^{Ik_x[(i-1)\Delta x]}\right\} \\
&= A^n e^{Iik_x x}\left(1 - \frac{u\Delta t}{2\Delta x}e^{Ik_x \Delta x} + \frac{u\Delta t}{2\Delta x}e^{-Ik_x \Delta x}\right)
\end{aligned} \tag{3.18}
$$

The amplification factor G can be expressed as follows:

$$
G = \frac{A^{n+1}}{A^n} = 1 - \frac{u\Delta t}{2\Delta x}e^{Ik_x \Delta x} + \frac{u\Delta t}{2\Delta x}e^{-Ik_x \Delta x} = 1 - I\frac{u\Delta t}{\Delta x}\sin(k_x \Delta x) \tag{3.19}
$$

which gives:

$$
|G| = \sqrt{1^2 + \left(\frac{u\Delta t}{\Delta x}\right)^2 \sin^2(k_x \Delta x)} > 1 \tag{3.20}
$$

It is evident that the value of the amplification factor is always greater than 1.0, indicating that the FTCS scheme is unconditionally unstable.

3.2.5 Schemes for convection equation

The convection–diffusion equation for pollution concentration C will be used as an example to introduce some classical difference schemes and their computational properties. The uniform grid is adopted. In this section, the schemes for linear convection equation is introduced first.

3.2.5.1 Implicit methods (Laasonen method, Crank–Nicolson method, and ADI method)

After realizing the unconditionally unstable problem of FTCS for a pure convection problem, an alternative idea is to include the information of the current time step $n + 1$ in the calculation of the spatial derivative as follows:

$$
\frac{C_i^{n+1} - C_i^n}{\Delta t} + u\left[(1-\gamma)\left(\frac{C_{i+1}^n - C_{i-1}^n}{2\Delta x}\right) + \gamma\left(\frac{C_{i+1}^{n+1} - C_{i-1}^{n+1}}{2\Delta x}\right)\right] = 0 \tag{3.21}
$$

where γ is the weighting factor, which varies from 0 to 1. When $\gamma = 0$, the aforementioned format returns to the FTCS scheme, and when $\gamma \neq 0$, this method is called the implicit method because the variables at $n + 1$ time step will be updated by solving a set of linear equations iteratively.

Laasonen method and Crank–Nicolson method: When $\gamma = 1$, the FD scheme is equivalent to backward time difference scheme, this format is also known as the *Laasonen method*. According to the heuristic stability analysis introduced earlier, this scheme is first-order accurate but the coefficient corresponding to the diffusion term is always positive, so it can contribute to suppress the numerical oscillation, and this scheme is stable. When $\gamma = 1/2$, it is equivalent to using the information between n step and $n + 1$ step to solve the spatial derivative; therefore, the scheme is of second-order accuracy in both time and space. The first term of the truncation error is the dispersion term of the third derivative, and the second term is the diffusion term of the fourth derivative. It is unconditionally stable based on the von Neumann stability analysis. However, because the diffusion term has higher order than the dispersion term, the suppression effect of dispersion error is not strong enough to remove numerical oscillation locally. This scheme is called the *Crank–Nicolson* method. Similar to the *Laasonen method*, it is unconditionally stable, and both schemes need to solve matrix.

Alternating direction implicit (ADI) method: The ADI method applies the time-splitting implicit method for multiple dimensional problems. It is known that when the implicit method is used, a matrix needs to be solved. For 1D problem, the bandwidth of the matrix is 3, which can be solved quickly. But for 2D or 3D problem, the matrix becomes large and sparse. To improve the calculation efficiency, the numerical solution can be carried out in each direction, step by step, after the ADI is applied to all directions alternately within one time step.

3.2.5.2 Explicit central-space methods (Lax method and Lax–Wendroff method)

Recall that the explicit FTCS has been proven to be an unconditionally unstable scheme. In this section, we will discuss how to stabilize the scheme by introducing different modifications without changing the original spatial center difference scheme of FTCS.

Lax method: The Lax method revises the FTCS by replacing the original value at the previous time-step center node with its spatial average from its left and right neighborhood nodes:

$$C_i^{n+1} = \frac{1}{2}\left(C_{i+1}^n + C_{i-1}^n\right) - u\frac{\Delta t}{2\Delta x}\left(C_{i+1}^n - C_{i-1}^n\right) \tag{3.22}$$

According to the heuristic stability analysis, we know that:

$$
\begin{aligned}
C_i^{n+1} &= \frac{1}{2}\left(C_{i+1}^n + C_{i-1}^n\right) - u\frac{\Delta t}{2\Delta x}\left(C_{i+1}^n - C_{i-1}^n\right) \\
&\overset{\text{equivalent}}{\leftrightarrow} \left(\frac{\partial C}{\partial t}\right)_i^n + u\left(\frac{\partial C}{\partial x}\right)_i^n = TE = \frac{1}{2\Delta t}\left(\Delta x^2 - \Delta t^2 u^2\right)\left(\frac{\partial^2 C}{\partial t^2}\right)_i^n \\
&\quad + O\left(\Delta x^2\right)
\end{aligned}
\tag{3.23}
$$

This scheme has the TE of $O(\Delta t,(\Delta x)^2)$ with first-order accuracy. Its stability condition can be easily determined as follows:

$$\Delta x^2 - \Delta t^2 u^2 \geq 0 \Leftrightarrow C_r = u\Delta t / \Delta x \leq 1 \qquad (3.24)$$

where C_r is called the Courant number and this condition is called Courant–Friedrichs–Lewy (CFL) condition, which is applicable to most of explicit FD schemes. It requires the time step to be small enough so that any numerical perturbations in a time step can only propagate to its adjacent grids. One of the drawbacks of this method is that it is too dissipative due to the first-order TE in time.

Lax–Wendroff method: Lax–Wendroff scheme is constructed to improve the accuracy of the Lax explicit scheme. An artificial positive diffusion term is introduced to balance the leading order truncation error of negative diffusion in FTCS. This classical scheme has the following form:

$$\frac{C_i^{n+1} - C_i^n}{\Delta t} + u\frac{C_{i+1}^n - C_{i-1}^n}{2\Delta x} = \frac{u^2\Delta t}{2\Delta x^2}\left(C_{i+1}^n - 2C_i^n + C_{i-1}^n\right) \qquad (3.25)$$

This scheme has the second-order accuracy in both time and space and is stable when $C_r \leq 1$.

3.2.5.3 Upwind schemes

To overcome the instability caused by FTCS, another approach is to keep the forward-time difference but to give up the central difference of the space derivative. One of the most common used schemes for this purpose is the first-order upwind scheme, for example:

$$\frac{C_i^{n+1} - C_i^n}{\Delta t} + u\left[(1-\gamma)\left(\frac{C_{i+1}^n - C_i^n}{\Delta x}\right) + \gamma\left(\frac{C_i^n - C_{i-1}^n}{\Delta x}\right)\right] = 0$$
$$\text{where } \gamma = \begin{cases} 0 & u \geq 0 \\ 1 & u < 0 \end{cases} \qquad (3.26)$$

In this scheme, when u is positive, the backward difference is used, and when u is negative, the forward difference is used. This scheme has first-order accuracy in both time and space but can suppress the numerical oscillation near the sharp front. The stability condition for this scheme is $C_r \leq 1$.

In order to improve the accuracy of the upwind scheme, some improved upwind schemes are proposed, that is, the QUICK (Quadratic Upwind Interpolation of Convective Kinematics) scheme (Leonard, 1979; Hayase et al., 1992) and the QUICKEST (Quadratic Upstream Interpolation for Convective Kinematics with Estimated Streaming Terms) scheme (Leonard,

1988). The *method of characteristics* can also be regarded as a special upwind format.

3.2.5.4 Oscillation control methods

Is there a scheme that is both high-order accuracy (second order or higher) and good at suppressing numerical oscillations? Such a scheme is necessary to capture shock motion. The commonly used method is to introduce a lower order accuracy scheme (e.g., first-order upwind or Lax method) or numerical diffusion at the beginning of numerical oscillation in a higher-order scheme, so that the overall numerical accuracy is still high-order, but the numerical oscillation can be controlled. *ENO (essentially nonoscillatory) schemes* and it extended scheme *WENO (weighted ENO)* (Liu et al., 1994) scheme can meet this requirement. Such schemes can automatically select the appropriate node to approximate the local derivative according to the smoothness of the function to ensure the high-order accuracy of the scheme in the smooth area and suppress the oscillation near the shock front.

In addition, Flux Limiter can be used to eliminate oscillations near the shock front. The limitation term must meet certain conservation principles. The commonly used limiters are van Leer, *ULTIMATE* (Universal Limiter for Transport Interpolation Modelling of the Advective Transport Equation), and Chakravarthy – Osher limitation, and so on.

3.2.5.5 Time-splitting methods

In this section, we will briefly introduce the classical time derivative difference schemes. Because most of these schemes split the solution of time derivative into two or more steps, we often refer to these schemes as time splitting methods.

MacCormack method: The MacCormack method is one of the simplest time splitting methods that divides the solution procedure into two time substeps:

$$\bar{C}_i^{n+1} = C_i^n - u\frac{\Delta t}{\Delta x}\left(C_{i+1}^n - C_i^n\right)$$

$$C_i^{n+1} = C_i^n - u\frac{\Delta t}{2\Delta x}\left(C_{i+1}^n - C_i^n + \bar{C}_i^{n+1} - \bar{C}_{i-1}^{n+1}\right) \tag{3.27}$$

The first step is called the predictor step, in which the tentative value of the variable $\bar{C}_i^{n+1}$ is obtained on basis of the information of n step. The second step is the corrector step, during which the tentative value is corrected and the final value is obtained. The scheme is simple and has high order of accuracy (second order in both space and time), but it has a drawback that oscillations may appear near a sharp front, similar to the Lax–Wendroff and Crank–Nicolson methods introduced earlier.

We can also use other methods to establish high-order accurate difference scheme for time derivative term. According to our understanding of the difference scheme, if we introduce more time nodes, such as n–1 step and even n–2 step information, it is possible to establish a difference scheme with higher order of accuracy in time. For example, the *Adams–Bashforth–Moulton method* is a higher-order accurate FD scheme in time by using two-level or multiple-level time splitting predictor-corrector method.

3.2.6 Schemes for diffusion equation

In environmental problems, we often need to solve convection–diffusion source–sink equations. The method introduced earlier can also be used to construct the FD schemes and to analyze the stability and accuracy of such equations. The additional terms are normally discretized by various orders of central differences, while the corresponding stability conditions can be obtained by the von Neumann analysis if the problem is linear. Now the FD scheme and the stability analysis for the following 1D diffusion equation is presented:

$$\frac{\partial C}{\partial t} = \varepsilon \frac{\partial^2 C}{\partial x^2} \tag{3.28}$$

The following FD scheme is used:

$$
\begin{aligned}
&\frac{C_i^{n+1} - C_i^n}{\Delta t} = \varepsilon \frac{C_{i+1}^n - 2C_i^n + C_{i-1}^n}{\Delta x^2} \\
&\overset{\text{equivalent}}{\longleftrightarrow} \left(\frac{\partial C}{\partial t} \right)_i^n = \varepsilon \left(\frac{\partial^2 C}{\partial x^2} \right)_i^n + TE = \varepsilon \left(\frac{\partial^2 C}{\partial x^2} \right)_i^n \\
&+ \frac{\varepsilon}{2} \left(\frac{\Delta x^2}{6} - \Delta t \varepsilon \right) \left(\frac{\partial^4 C}{\partial x^4} \right)_i^n + O(\Delta x^3)
\end{aligned} \tag{3.29}
$$

It can be found that this scheme is of second-order accuracy and conditionally stable when the coefficient of the first term of TE is nonnegative, namely:

$$\frac{\Delta x^2}{6} - \Delta t \varepsilon \geq 0 \Rightarrow \Delta t \leq \frac{\Delta x^2}{6\varepsilon} \tag{3.30}$$

This stability condition is similar to the CFL condition required to solve the convection equation in an explicit scheme, which require time steps to be small enough so that no information passage can exceed a grid size in a time step through diffusion process.

When the FD scheme is used to solve the convection–diffusion equation, the error of the low-order derivative (convection term) may be the same as that of the higher-order derivative (diffusion term). In this case, both the stability condition and the solution accuracy will be affected. Taking the 1D convection–diffusion equation as an example, we use the upwind scheme to solve the convection term and the central difference scheme to solve the diffusion term. Through error analysis, we obtain the following truncation error:

$$\frac{\partial C}{\partial t} + u\frac{\partial C}{\partial x} - \varepsilon\frac{\partial^2 C}{\partial x^2} = TE = \frac{u}{2}(\Delta x - u\Delta t)\frac{\partial^2 C}{\partial x^2} + O(\Delta t^2, \Delta x^2) \qquad (3.31)$$

We find that the first term of TE has the same expression as the diffusion, except that the corresponding diffusion coefficient is $D_N = \frac{u}{2}(\Delta x - u\Delta t)$. In general, in order to ensure the stability, both the convection stability and the diffusion stability are required to be satisfied. When the diffusion process dominates, it requires $D_N \ll \varepsilon$ to ensure that the FD scheme represents the correct physical process.

The source and sink terms: The water quality equation always has a source or sink term that represents the process in which substances are changed by various biochemical processes. In numerical calculation, since the item is generally related to the substance concentration itself and can be described by algebraic functions, it is not necessary to establish a separate difference scheme for it. However, we need to determine which step of the sink term information is used in the FD scheme. If the n step information is used, the similar stability condition is required, that is, if the time step is too large, the calculated concentration may change from positive to negative within one time step. A better approach is to use the $n + 1$ step information, which is similar to the implicit scheme of the FD scheme. Take the linear decay equation as an example:

$$\frac{\partial C}{\partial t} = -rC \qquad (3.32)$$

The implicit FD scheme is expressed as follows:

$$\frac{C_i^{n+1} - C_i^n}{\Delta t} = -rC_i^{n+1} \Rightarrow C_i^{n+1} = \frac{C_i^n}{1 + \Delta tr} \qquad (3.33)$$

Interested readers are invited to use the n step information on the right-hand side instead to compare the difference between the two approaches.

Example 3.1 The numerical simulation of a one-dimensional turbulent advection–diffusion system.

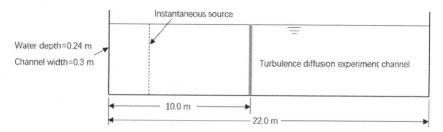

Figure 3.3 Sketch of a turbulence diffusion channel.

Problem: Figure 3.3 shows a channel of hydraulic lab. At the beginning of the experiment, the tracer (0.18 kg) is input as 'instantaneous source' at distance of 10.0 m from the left boundary. The direction of the flow from left to right with velocity $u = 0.1$ m/s. The turbulence diffusion coefficient in the channel is $\varepsilon = 0.03$ m²/s.

1. Derive the mathematical model for simulating the turbulent diffusion problem and write appropriate difference scheme for the governing equation.
2. Compare the simulation results with the analytical solution, and discuss the results.

Solution: 1. Because the tracer is uniformly distributed at channel section, this problem can be considered as one-dimensional advection–turbulence diffusion problem, for which the governing equation can be written as follows:

$$\frac{\partial C}{\partial t} + u\frac{\partial C}{\partial x} - \varepsilon\frac{\partial^2 C}{\partial x^2} = 0$$

As Figure 3.3 shows, the left and right boundaries of the channel are located at $x = 0$ m and $x = 22.0$ m. The open boundary condition is used at the left and right boundaries:

$$\frac{\partial C}{\partial x} = 0 \text{ at } x = 0 \text{ m and } x = 22.0 \text{ m}$$

The initial condition can be written as follows:

$C(x,0) = M$, when $x = 10.0$ m
$C(x,0) = 0$, when $x \neq 10.0$ m

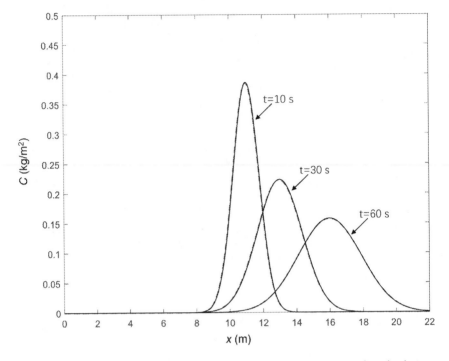

Figure 3.4 Comparisons of the numerical results (dashed line) and the analytical solutions (solid line) for an advection–diffusion problem.

The upwind scheme is used to discretize the convection term, and the central difference scheme is used to discretize the diffusion term:

$$\frac{\partial C}{\partial t} + u\frac{\partial C}{\partial x} - \varepsilon\frac{\partial^2 C}{\partial x^2} = \frac{C_i^{n+1} - C_i^n}{\Delta t} + u\left[(1-\gamma)\left(\frac{C_{i+1}^n - C_i^n}{\Delta x}\right) + \gamma\left(\frac{C_i^n - C_{i-1}^n}{\Delta x}\right)\right]$$
$$- \varepsilon\frac{C_{i+1}^n - 2C_i^n + C_{i-1}^n}{\Delta x^2} \qquad \text{where } \gamma = \begin{cases} 0 & u \geq 0 \\ 1 & u < 0 \end{cases}$$

2. In a one-dimensional system, the corresponding mass unit is kg/m²; therefore, in this problem, $M = 0.18/0.24 = 0.75$ kg/m².

In the numerical simulation, the grid size is $\Delta x = 0.02$ m, $\Delta t = 0.01$ s. Figure 3.4 shows comparisons of the simulated concentration with the analytical solution (Eqn. (2.57)) along the channel.

3.3   FINITE ELEMENT METHOD

3.3.I   Background of finite element method

The FEM is an effective method to solve mathematical problems. Different from the FDM that approximates the derivative term of a PDE, FEM approximates the solution of the PDE directly. In FEM, the computation domain is divided into several nonoverlapping but interconnected units, for each of which a basis function is selected to approximate the true solutions. The basis function for the whole calculation domain can be considered as the superposition of the basis function of each unit, and then, the solution for the computation domain can be taken as the superposition of approximate solutions on all units. In general, the FEM can flexibly use various shapes of units and grids, so it is suitable to simulate the problems with complex boundary shapes. The FEM is more popularly used in solid mechanics than fluid mechanics, with the exception of groundwater simulation.

3.3.2   Domain discretization

The first step of the FE approach is to discretize the computational domain into several computing units based on the shape of the computation region. In addition to numbering the calculation elements and the nodes and determining the relationship between them, the position coordinates of the nodes should also be established. Take 1D problem, for example, assuming the domain length is L and the domain is discretized into a finite number of N uniform elements, the length of each element is $\Delta x = L / N$ and each element has N_n nodes (Figure 3.5). For 1D problems, the element can only be a line segment; for 2D problems, the commonly used elements shapes are triangles and squares; and for 3D problems, the shape of the element can take different forms (Figure 3.6).

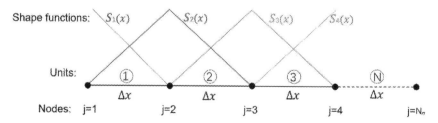

Figure 3.5 Finite element method for one-dimensional problem.

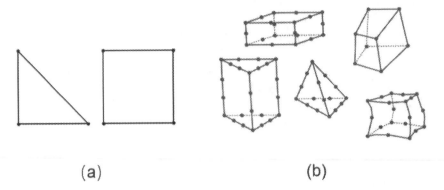

(a)   (b)

Figure 3.6 The commonly used elements in (a) 2D and (b) 3D finite element method.

3.3.3 Shape function

By dividing the entire domain into N elements, the total node number is N_n. We now try to approximate the true solution of ϕ (x,t) in the entire domain by the approximate function $\tilde{\phi}(x,t)$:

$$\phi(x,t) \approx \tilde{\phi}(x,t) = \sum_{j=1}^{N_n} S_j(x)\phi_j(t) \tag{3.34}$$

where $S_j(x)$ is the shape function corresponding to nodal point j and $\phi_j(t)$ is the approximate function of the nodal point j at time t to be solved. Because ϕ_j is an approximation of the true solution on node j, the shape function $S_j(x)$ will be 1 on node j but linearly reduces to 0 to its nearest neighborhood nodes. This gives the design principle of the shape function. In practice, we can have different shape functions that satisfy this principle, but the simplest functions are obviously linear (i.e., first-order polynomials). By using the piecewise linear function, we can easily construct the shape function:

$$S_j(x) = \begin{cases} \dfrac{x-(j-2)\Delta x}{\Delta x} & \text{for } (j-2)\Delta x \le x \le (j-1)\Delta x \\ 1 - \dfrac{x-(j-1)\Delta x}{\Delta x} & \text{for } (j-1)\Delta x \le x \le j\Delta x \\ 0 & \text{elsewhere} \end{cases} \tag{3.35}$$

3.3.4 Error minimization

If the $\tilde{\phi}(x,t)$ is the approximation solution to $\phi(x,t)$, when substituting $\tilde{\phi}(x,t)$ into the original *PDE*, a small error (or residual) ε_r will be resulted:

$$\frac{\partial \tilde{\phi}}{\partial t} + u \frac{\partial \tilde{\phi}}{\partial x} = \varepsilon_r \tag{3.36}$$

Obviously, we are unable to make the error be zero in the entire domain, but we can minimize the accumulated error in the domain. To achieve this, we introduce N_w linearly independent weighting functions W_i and require the weight of residual error integral to be zero:

$$\int_V W_i \varepsilon_r dV = \int_V W_i \left(\frac{\partial \tilde{\phi}}{\partial t} + u \frac{\partial \tilde{\phi}}{\partial x} \right) dV = 0 \tag{3.37}$$

where W_i is the weighting function. During the calculation, similar to the shape function, different weighting functions can be introduced in the computation, resulting in different weighted residuals.

Collocation method: If the weighting function is the Delta function, this method is called collocation method:

$$W_i = \delta(\vec{x} - \vec{x}_i) \tag{3.38}$$

For all nodes, Eqn. (3.37) becomes:

$$\int_V W_i \varepsilon_r dV = \int_V \delta(\vec{x} - \vec{x}_i) \varepsilon_r dV = \varepsilon_i = 0 \tag{3.39}$$

In this method, the residuals are made to be zero at all nodes but not on the elements. Although this method seems simple and can guarantee that the error of the node is zero, the instability may occur in practical computations.

Galerkin method: If the weight function is the same as the shape function, that is, $W_i = S_i$, the method is called the Galerkin method:

$$\int_V W_i \varepsilon_r dV = \int_V S_i(x) \left(\frac{\partial \sum_{j=1}^{N_n} S_j(x)\phi_j(t)}{\partial t} + u \frac{\partial \sum_{j=1}^{N_n} S_j(x)\phi_j(t)}{\partial x} \right) dV = 0 \tag{3.40}$$

In this case, when N_n approaches infinity, the error residual in the whole domain approaches zero, which not only guarantees the consistency of the finite element solution theoretically but also guarantees the minimization of the overall error. This is the most commonly used FEM, which is stable but requires large amount of computation.

Least-square method: Another possible function is the error residue itself. This method is not able to guarantee the integral residual error to be zero, but it can minimize the error by making the derivative with respect to S_i to be zero:

$$\frac{\partial}{\partial W_i} \int_V \varepsilon_r^2 dV = 2 \int_V \frac{\partial \varepsilon_r}{\partial W_i} \varepsilon_r dV = 0 \tag{3.41}$$

This method is called the least-square (LS) method. It is equivalent to choosing the weighting function as $\partial \varepsilon / \partial W_i$ in Eqn. (3.37).

3.3.5   Matrix assembly

After obtaining the finite element equation for each element, all of the finite element equations in the computation region can be used to form a global matrix. When the shape function $S_i(x)$ is of a simple polynomial and we express the time derivative $\partial \phi_j(t) / \partial t$ in Eqn. (3.40) as the FD scheme $\left(\phi_j^{n+1} - \phi_j^n\right) / \Delta t$, the following matrix can be obtained:

$$A_{ij} \phi_j^{n+1} = f_i \left(\phi^n\right) \tag{3.42}$$

where A_{ij} is the matrix of $N_n \times N_n$, ϕ_j is the unknown function to be solved at all N_n nodes, and f_i is the forcing vector of size N_n that incorporates all nodal conditions at previous time step. Both A_{ij} and f_i are determined according to the characteristics of the PDE to be solved and the choice of FE scheme.

In practical computation, the shape function S_i is nonzero only in the elements adjacent to node i; therefore, only the local element integration is needed when we carry out the global integration. For a 1D problem, Eqn. (3.40) can be rewritten as follows:

$$\int_V S_i(x) \left(\frac{\partial \sum_{j=1}^{N_n} S_j(x)\phi_j(t)}{\partial t} + u \frac{\partial \sum_{j=1}^{N_n} S_j(x)\phi_j(t)}{\partial x} \right) dV$$

$$= \sum_{e=i-1}^{i} \int_{V^{(e)}} S_i(x) \left(\frac{\partial \sum_{m=1}^{N_e} S_m^{(e)}(x)\phi_m^{(e)}(t)}{\partial t} + u \frac{\partial \sum_{m=1}^{N_e} S_m^{(e)}(x)\phi_m^{(e)}(t)}{\partial x} \right) dV = 0 \tag{3.43}$$

where $V(e)$ is the integration domain for the eth element and $s_m^{(e)}(x)\phi_m^{(e)}(t)$ is equivalent to $S_{e+m-1}(x)\,\phi_{e+m-1}(t)$. They are defined on the basis of element information and global nodal information, respectively. For particular i, Eqn. (3.43) can be rewritten as follows:

$$
\int_{(i-2)\Delta x}^{(i-1)\Delta x} s_2^{(i-1)}(x) \left[\frac{\partial \sum_{m=1}^{2} s_m^{(i-1)}(x)\phi_m^{(i-1)}(t)}{\partial t} + u \frac{\partial \sum_{m=1}^{2} s_m^{(i-1)}(x)\phi_m^{(i-1)}(t)}{\partial x} \right] dx
$$

$$
+ \int_{(i-1)\Delta x}^{i\Delta x} s_1^{(i)}(x) \left[\frac{\partial \sum_{m=1}^{2} s_m^{(i)}(x)\phi_m^{(i)}(t)}{\partial t} + u \frac{\partial \sum_{m=1}^{2} s_m^{(i)}(x)\phi_m^{(i)}(t)}{\partial x} \right] dx = 0
$$

(3.44)

The global integration can be expressed by individual element integration; therefore, the integration can be performed on each element with reference to local coordinate (i.e., from 0 to Δx). In this way, the integration process can be greatly simplified if the geometric properties of all the elements are the same. By using $s_m^{(e)}(x) = S_{m+e-1}$ and Eqn. (3.35), Eqn. (3.44) can be rewritten as follows:

$$
\int_0^{\Delta x} \frac{x}{\Delta x} \left[\frac{\partial \left[\left(1 - \frac{x}{\Delta x}\right)\phi_1^{(k-1)}(t) + \frac{x}{\Delta x}\phi_2^{(k-1)} \right]}{\partial t} + u \frac{\partial \left[\left(1 - \frac{x}{\Delta x}\right)\phi_1^{(k-1)}(t) + \frac{x}{\Delta x}\phi_2^{(k-1)} \right]}{\partial x} \right] dx
$$

$$
+ \int_0^{\Delta x} \left(1 - \frac{x}{\Delta x}\right) \left[\frac{\partial \left[\left(1 - \frac{x}{\Delta x}\right)\phi_1^{(k)}(t) + \frac{x}{\Delta x}\phi_2^{(k)} \right]}{\partial t} + u \frac{\partial \sum_{m=1}^{N_e} \left[\left(1 - \frac{x}{\Delta x}\right)\phi_1^{(k)}(t) + \frac{x}{\Delta x}\phi_2^{(k)} \right]}{\partial x} \right] dx = 0
$$

(3.45)

Once the integration for each element is obtained, it can be assembled to obtain the coefficients A_{ij} associated with each $\phi_m(e)$.

3.3.6 Solution to linear system of equations

After obtaining the matrix coefficients and the right-hand side terms, the unknowns can be solved by an appropriate matrix solver. The techniques of matrix solving are introduced in linear algebra and numerical analysis courses. Here, we give a brief introduction to some commonly used methods.

The linear equations can be expressed generally in the following form:

$$Ax = b \qquad (3.46)$$

where A is a matrix of $N_n \times N_n$ and N_n is the total node number. The coefficient of the matrix is a_{ij}. The solution of the matrix can be obtained as follows:

$$x = A^{-1}b \qquad (3.47)$$

where A^{-1} is the inverse matrix of A.

When A is a full matrix, the most commonly used method is Gaussian elimination. This method can provide the exact solution of the matrix, but the calculation is time consuming for a large matrix. The formation of a full matrix only occurs when one node is intrinsically related to all other nodes in the domain. In the process of numerical calculation, both time and space are discretized, and we usually only need to use adjacent nodes for the calculation of a node in a time step, and then, the matrix we obtain becomes a sparse matrix, that is, most of the coefficients of the matrix are zero. For sparse matrices, we can adopt more efficient methods in both matrix storage and matrix solution, so as to obtain the numerical solution of the matrix more efficiently. The common methods to solve sparse matrix include the Gauss-Seidel method, the successive over-relaxation (SOR) method, and the conjugate gradient (CG) method.

3.3.6.1 Gauss–Seidel method and Jacobi method

The Gauss–Seidel method is a basic method to solve linear equations by the iterative method. This method uses the following iterative equations to approximate the numerical solutions of the matrix system:

$$x_i^{(k+1)} = \frac{1}{a_{ii}}\left(b_i - \sum_{j=1}^{i-1} a_{ij} x_j^{(k+1)} - \sum_{j=i+1}^{N_n} a_{ij} x_j^{(k)} \right) \qquad (3.48)$$

The aforementioned calculations are performed only on nonzero coefficients. Once the solution on a node is obtained, it will participate in the calculation of the next node, so there is no need to store the information of the previous step, thus saving the storage space. The Gauss–Seidel method can be considered as an improvement of the traditional and slower Jacobi method, which always uses the information of the previous iteration step to calculate the value of the next iteration step, that is, replacing $x_j^{(k+1)}$ on the right side of the equation by $x_j^{(k)}$.

3.3.6.2 Successive over-relaxation method

Compared with the Gauss–Seidel method, SOR method increases the convergence rate by multiplying the acceleration coefficient. The scheme is as follows:

$$x_i^{(k+1)} = \left(1 - \omega\right)x_i^{(k)} + \frac{\omega}{a_{ii}}\left(b_i - \sum_{j=1}^{i-1}a_{ij}x_j^{(k+1)} - \sum_{j=i+1}^{N_n}a_{ij}x_j^{(k)}\right) \tag{3.49}$$

For the symmetric positive definite matrix, it can be proved that the matrix solution is convergence when $0 < \omega < 2$ and is equivalent to the Gauss–Seidel method when $\omega = 1$. When $\omega > 1$, the method is called the SOR method, with the optimal ω value being determined by the characteristics of matrix.

3.3.6.3 Conjugate gradient method

For a symmetric sparse positive definite matrix, there exists a more efficient algorithm named the CG method. Two nonzero vectors u and v are called conjugates for matrix A if they satisfy the following conditions:

$$u^T A v = 0 \tag{3.50}$$

In the CG method, we first expand the matrix solution in series in n orthogonal spaces:

$$x = \alpha_1 p_1 + \ldots + \alpha_n p_n = \sum_{k=1}^{n}\alpha_k p_k \tag{3.51}$$

where p_k is a series of conjugate vectors, representing orthogonal vector spaces. The coefficient can be expressed as follows:

$$\alpha_k = \frac{p_k^T b}{p_k^T A p_k} \tag{3.52}$$

The main challenge is how to determine p_k. Because of the series expansion, we do not need to find all p_k but only determine the first few terms of p_k to approximate the real solution accurately.

In the iteration, assuming that any initial approximate solution is x_0, then $p_k = p_p = b - Ax_0$, and the first calculation adopting:

$$p_{k+1} = r_k - \frac{p_k^T A r_k}{p_k^T A p_k} p_k \tag{3.53}$$

where $r_k = b - Ax_k$ is the residual error.

When the residual error is less than a specific small value $\|r_k\| < \varepsilon$, the iteration stops. The theoretical basis of CG is as follows: the solution of $Ax = b$ is equivalent to the minimization of $f(x) = x^T A x / 2 - b^T x$, which is equivalent to the solution of $\frac{\partial f(x)}{\partial x} = 0$. The solution of p_k by the CG method is equivalent to the solution of the gradient of $f(x)$ at $x = x_k$, and the subsequent process is to find the conjugate vector of the gradient.

3.4 NUMERICAL METHODS IN QUAL-2K: A RIVER AND STREAM WATER QUALITY MODEL

QUAL-2K is the latest model in QUAL series of the river water quality model. QUAL-2K is suitable for simulating hydrological and water quality conditions of a small river. It is a simple one-dimensional model that simulates basic stream transport and mixing processes and allows users to specify many of the kinetic parameters on a reach-specific basis (Chapra et al., 2008).

Based on steady flow assumption, QUAL-2K can simulate the transport of various pollutants and water quality parameters in river water as a function of space and time, including BOD, DO, temperature, algae, chlorophyll, ammonia nitrogen, nitrite, nitrate, dissolved phosphorus, coliform bacteria, and so on.

The basic mass transport equations solved by the QUAL-2K model is the one-dimensional advection–dispersion equation:

$$\frac{\partial C}{\partial t} = \frac{\partial \left(A_x E_x \frac{\partial C}{\partial x} \right)}{A_x \partial x} - \frac{\partial (A_x u C)}{A_x \partial x} + \frac{dC}{dt} + \frac{S_c}{A_x \Delta x} \tag{3.54}$$

where A_x is the river cross-sectional area and S_c is the external source and sink for pollutant. The governing equation includes the advection, diffusion, dilution, and sources and sinks. For pollutant concentration C, the term dC/dt represents the pollutant growth and decay, for which the detailed expressions are presented in Chapra et al. (2008).

Figure 3.7 shows the grid discretization of a river, with the main tributary of the river being divided into different river reaches, and each reach is divided into several calculation elements. The river system in Figure 3.7 has 4 river reaches and 29 elements.

Each calculation element in QUAL-2K can be regarded as a completely mixed unit, and the outflow is a constant Q_i (see Figure 3.8). The flow quality can be calculated by

$$Q_i = Q_{i-1} \pm Q_{x,i} \tag{3.55}$$

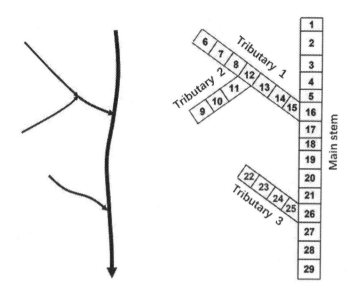

(a) A river with tributaries   (b) Q2K reach representation

Figure 3.7 Schematic diagram of the QUAL-2K river simulation network cutting: (a) a river with tributes and (b) Q2K reach representation.

Source: Chapra et al., 2008

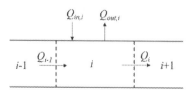

Figure 3.8 Schematic diagram of flow quality conservation in the QUAL-2K model calculation element.

Source: Chapra et al., 2008

where Q_{i-1} is the inflow from the upstream element, $Q_{in, i}$ is the lateral entering flow, and $Q_{out, i}$ is the lateral discharging flow.

The backward implicit FDM is used to discretize the governing Eqn. (3.50). The advection and diffusion terms are first solved with respect to x, which gives:

$$\frac{\partial C_i}{\partial t} = \frac{\left(A_x E_x \frac{\partial C}{\partial x}\right)_i - \left(A_x E_x \frac{\partial C}{\partial x}\right)_{i-1}}{V_i} - \frac{(A_x uC)_i - (A_x uC)_{i-1}}{V_i} \\ + \frac{dC_i}{dt} + \frac{S_{ci}}{V_i}$$

(3.56)

where $V_i = A_{x_i} \Delta x_i$.

Second, the time derivative and the spatial derivative of the diffusion terms in FD are written as follows:

$$\frac{C_i^{n+1} - C_i^n}{\Delta t} = \frac{\left[(A_x E_x)_i C_{i+1}^{n+1} - (A_x E_x)_i C_i^{n+1}\right]}{V_i \Delta x_i} \\ - \frac{\left[(A_x E_x)_{i-1} C_i^{n+1} - (A_x E_x)_{i-1} C_{i-1}^{n+1}\right]}{V_i \Delta x_i} \\ - \left(\frac{Q_i C_i^{n+1} - Q_{i-1} C_{i-1}^{n+1}}{V_i}\right) + r_i C_i^{n+1} + p_i + \frac{S_{ci}}{V_i}$$

(3.57)

where $Q_i = A_{x_i} u_i$ and the term dC_i/dt is expressed as follows:

$$\frac{dC_i}{dt} = r_i C_i^{n+1} + p_i$$

(3.58)

where r_i is the first-order rate constant and P_i is the internal pollutant source and sink.

Rearranging Eqn. (3.53) in terms of the coefficient of C_{i-1}^{n+1}, C_i^{n+1}, and C_{i+1}^{n+1}, we obtain:

$$a_i C_{i-1}^{n+1} + b_i C_i^{n+1} + c_i C_{i+1}^{n+1} = z_i$$

(3.59)

where

$$a_i = -(A_x E_x)_{i-1} \frac{\Delta t}{V_i \Delta x_i} - \frac{Q_{i-1} \Delta t}{V_i} \\ b_i = 1.0 + \left[(A_x E_x)_i + (A_x E_x)_{i-1}\right] \frac{\Delta t}{V_i \Delta x_i} + \frac{Q_i \Delta t}{V_i} - r_i \Delta t \\ c_i = -(A_x E_L)_i \frac{\Delta t}{V_i \Delta x_i} \\ z_i = c_i^n + p_i \Delta t + \frac{S_{ci} \Delta t}{V_i}$$

(3.60)

The value of a_i, b_i, c_i, and z_i are all known at time step n, and the C_{i-1}^{n+1}, C_i^{n+1}, and C_{i+1}^{n+1} terms are unknowns at time step $n + 1$.

In the case of a junction element with a tributary upstream element, the basic equation becomes:

$$a_i C_{i-1}^{n+1} + b_i C_i^{n+1} + c_i C_{i+1}^{n+1} + d_j C_j^{n+1} = z_i \qquad (3.61)$$

where $d_j = -\left(A_x E_x\right)_j \dfrac{\Delta t}{V_i \Delta x_i} - \dfrac{Q_j \Delta t}{V_i}$, j is the element upstream of junction element I, and C_j^{n+1} is the pollutant concentration in element j at time step $n + 1$. It can be observed that the term d_j is analogous to the a_i term. Both terms account for mass inputs from upstream due to dispersion and advection.

Equations (3.55) and (3.57) represent a set of simultaneous linear equation whose solution provides the value of C_i^{n+1} for all elements. By giving proper boundary and initial conditions, the solutions for Eqns. (3.55) and (3.57) can be obtained by solving the matrix. The application of the QUAL-2K model in water quality simulation in a river will be presented in Chapter 6.

3.5 NUMERICAL METHODS IN SUTRA: A GROUNDWATER FLOW MODEL

SUTRA is a computer model, which was developed by U.S. Geological Survey to simulate the fluid flow and solute or energy transport in subsurface systems (Voss, 1984). The original version of SUTRA was distributed in 1984. The revised version of SUTRA was distributed in 1990. The second revised version was distributed in 1997, which added the option to use Fortran 90 dynamic array dimensioning. It has been widely used for different types of groundwater flow analysis (Voss, 1999). The model SutraGUI was distributed in 2010. It is a flexible graphical user interface (GUI) support two-dimensional (2D) and three-dimensional (3D) simulation.

SUTRA can simulate the saturated/unsaturated, variable-density groundwater flow with solute transport by solving the groundwater flow and solute or energy transport equations (Voss and Provost, 2002). The water head, pollutant concentration, and temperature changes can be obtained from this model.

In SUTRA, the following fluid mass balance equation is used to simulate the transient propagation of the solute front as it moves radially away from the well:

$$\left(S_w \rho S_{op} + e\rho \frac{\partial S_w}{\partial p}\right)\frac{\partial p}{\partial t} + \left(eS_w \frac{\partial \rho}{\partial C}\right)\frac{\partial C}{\partial t} - \nabla \cdot \left[\left(\frac{k_p \rho}{\mu_m}\right)(\nabla p - \rho g)\right] = Q_p \quad (3.62)$$

where S_w is the water saturation, S_{op} is the specific pressure storativity, e is the porosity (volume of voids per total volume), k_p is the permeability, and Q_p is the fluid mass source.

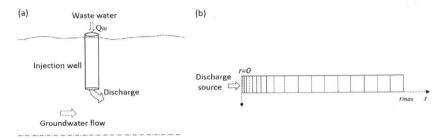

Figure 3.9 Schematic diagram of the SUTRA solute and energy transport.

The numerical methods used in SUTRA is based on a hybridization of finite-element and integrated finite-difference methods (Voss and Provost, 2002). In the balance equations, the terms that describe fluxes of fluid mass, solute mass, and energy are approximated by the standard finite-element method. The hybrid method is used to approximate the nonflux terms. We now use the example of groundwater contamination by polluted water from an inject well to illustrate how SUTRA handles a realistic problem. Figure 3.9(a) shows an example of a confined infinite aquifer, which contains an injection well. Assume the fluid injection rate is Q_{in}, with a solute concentration of C_s. Figure 3.9(b) shows the radial finite-element mesh for solute and energy transport.

As introduced in Section 3.3, the weighted residual can be written as follows:

$$\varepsilon_r = \left[S_w \rho S_{op} + e\rho \frac{\partial S_w}{\partial p} \right] \frac{\partial p}{\partial t} + \left(eS_w \frac{\partial \rho}{\partial C} \right) \frac{\partial C}{\partial t} - \nabla \cdot \left[\left(\frac{k_p \rho}{\mu_m} \right) \cdot (\nabla p - \rho g) \right]$$
$$- Q_p \approx 0 \qquad (3.63)$$

When the expression is approximated, the solution is no longer exactly equal to zero. A weighted residual ε_r formulation can be written as follows:

$$\int_V W_i \varepsilon_r dV = 0 \qquad (3.64)$$

The weighting function W_i is chosen to be either the basis function φ_i or the asymmetric weighting function w_i depending on the term of the equation. The integral terms in the fluid mass balance equation are approximated discretely and substituted in Eqn. (3.63).

The first term is an integral of the pressure derivative, and the weighting function is chosen to be the basis function:

$$\int_V \left[\left(S_w \rho S_{op} + e\rho \frac{\partial S_w}{\partial p} \right) \frac{\partial p}{\partial t} \right] \phi_i dV = \left(S_w \rho S_{op} + e\rho \frac{\partial S_w}{\partial p} \right)_i \frac{\partial p}{\partial t} V_i \qquad (3.65)$$

The second term of Eqn. (3.62) is also a time derivative, which is also approximated by:

$$\int_V \left[\left(eS_w \frac{\partial \rho}{\partial C}\right)\frac{\partial C}{\partial t}\right]\phi_i dV = \left(eS_w \frac{\partial \rho}{\partial C}\right)_i \frac{\partial C_i}{\partial t} V_i \qquad (3.66)$$

The third term in Eqn. (3.62) involves the divergence of fluid flux and is weighted with the asymmetric function ω_i. The asymmetry function is used for unsaturated flow problems only to ensure solution stability when the meshes are not fine enough to represent sharp saturation fronts, which gives:

$$\int_V \left\{\nabla \cdot \left[\left(\frac{k_p\rho}{\mu_m}\right)(\nabla p - \rho g)\right]\right\}\omega_i dV$$
$$= \int_V \left[\left(\frac{k_p\rho}{\mu_m}\right)\nabla \phi_j\right]\nabla \omega_i dV + \int_V \left[\left(\frac{k_p\rho}{\mu_m}\right)\rho g\right]\nabla \omega_i dV \qquad (3.67)$$

By introducing different weighting functions in the FE scheme, the following weighted residual relation for (3.62) is obtained in a 2D problem:

$$AF_i \frac{\partial p_i}{\partial t} + CF_i \frac{\partial C_i}{\partial t} + \sum_{j=1}^{N_p N_n} p_j(t)BF_{ij} = DF_i + Q_i \qquad (3.68)$$

where

$$AF_i = \left(S_w\rho S_{op} + e\rho\frac{\partial S_w}{\partial p}\right)_i V_i \qquad (3.69)$$

$$CF_i = \left(eS_w \frac{\partial \rho}{\partial C}\right)_i V_i \qquad (3.70)$$

$$BF_i = \int_V \left[\left(\frac{k_p\rho}{\mu_m}\right)\nabla \phi_j\right]\nabla \omega_i dV \qquad (3.71)$$

$$DF_i = \int_V \left[\left(\frac{k_p\rho}{\mu_m}\right)\rho g\right]\nabla \omega_i dV \qquad (3.72)$$

The only integrals requiring Gaussian integration is BF_i and DF_i.

The time derivatives in the spatially discretized and integrated equation are approximated by FDs. The pressure term is approximated by forward difference scheme, which gives:

$$\frac{\partial p_i}{\partial t} = \frac{p_i^{n+1} - p_i^n}{\Delta t} \qquad (3.73)$$

The time derivative of concentration term is approximated by backward difference scheme, which gives:

$$\frac{\partial C_i}{\partial t} = \frac{C_i^n - C_i^{n-1}}{\Delta t} \tag{3.74}$$

The form of the discretized fluid mass balance implemented in SUTRA is as follows:

$$\sum_{j=1}^{N_n} \left[\frac{AF_i\, \delta_{ij}}{\Delta t} + BF_{ij}^{n+1} \right] p_j^{n+1}$$
$$= Q_i^{n+1} + DF_i^{n+1} + \left(\frac{AF_i^{n+1}}{\Delta t} \right) p_i^n + CF_i^{n+1} \left(\frac{\partial C_i}{\partial t} \right)^n \tag{3.75}$$

where δ_{ij} is the Kronecker delta:

$$\delta_{ij} = \begin{cases} 0, & i \neq j \\ 1, & i = j \end{cases} \tag{3.76}$$

A global matrix can be formed on the basis of the finite element equation (refer to Section 3.3.5), and then, the unknowns can be solved by an appropriate matrix solver (refer to Section 3.3.6). The detailed numerical approximation of SUTRA fluid mass balance equation can be found in Voss and Provost (2002).

EXERCISES

1. Consider the following PDE:

$$\frac{\partial f}{\partial t} + b_1 \frac{\partial f}{\partial x} + b_2 \frac{\partial^3 f}{\partial x^3} = b_3 \frac{\partial^2 f}{\partial x^2}$$

The initial condition is:

$$f_{t=0} = \begin{cases} 0 & 0 \leq x < 5 \\ 1 & 5 \leq x < 12 \\ 0 & 12 \leq x < 20 \end{cases}$$

The boundary condition is:

$$f_{x=0}(t) = 0 \text{ and } f_{x=20}(t) = 0$$

(a) Write appropriate FD scheme for the following cases:

   1) $b1 = 1, b2 = 0, b3 = 0$; use explicit scheme.
   2) $b1 = 1, b2 = 0, b3 = 0$; use implicit scheme.
   3) $b_1 = 1, b_2 = 0, b_3 = 1$.
   4) $b_1 = 0, b_2 = 1, b_3 = 0$.
   5) $b_1 = 1, b_2 = 1, b_3 = 0$.

(b) Use the von Neumann method to analyze the numerical stability of the difference schemes and compare it with heuristic stability analysis method.

(c) Write a computer program based on the FD scheme used (i.e., FORTRAN, C, C++, BASIC, MATLAB®, etc.).

(d) Determine appropriate time step and grid size; run the computer program until $t = 30$.

(e) Analyze and discuss the calculated results. Compare the calculated results with the analytical results, if possible. Change time step and grid size to test the stability and grid convergence.

2. Solve the following one-dimension linear convection equation by using the Lax–Wendroff method:

$$\frac{\partial \phi}{\partial t} + b \frac{\partial \phi}{\partial x} = 0$$

(a) Use the heuristic stability analysis method to find the truncation error of this difference scheme.

(b) Discuss the stability and accuracy of the method according to the truncation error.

(c) Figure 3.10 shows the distribution of variable at the initial moment. Try to draw the distributions of variable calculated by the aforementioned difference scheme after a period of time (assume $C > 0$ and neglect boundary effects).

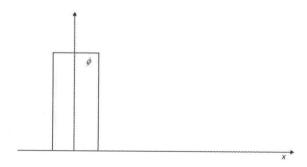

Figure 3.10 Initial distribution of the variable.

REFERENCES

Chapra, S.C., Pelletier, G.J. and Tao, H. (2008). QUAL2K: A modeling framework for simulating river and stream water quality. *Version 2.11: Documentation and Users Manual. Civil and Environmental Engineering Dept.* Tufts University, Medford, MA, Steven.

Hayase, T., Humphrey, J.A.C. and Greif, R. (1992). A consistently formulated QUICK scheme for fast and stable convergence using finite-volume iterative calculation procedures. *Journal of Computational Physics*, 98(1), pp. 108–118.

Leonard, B.P. (1979). A stable and accurate convective modelling procedure based on quadratic upstream interpolation. *Computer Methods in Applied Mechanics and Engineering*, 19(1), pp. 59–98.

Leonard, B.P. (1988). Elliptic systems: Finite-difference method IV. In Minkowycz, W.J., Sparrow, E.M., Schneider, G.E. and Pletcher, R.H. (eds.), *Handbook of Numerical Heat Transfer*. Wiley, New York, pp. 347–378.

Liu, X.D., Osher, S. and Chan, T. (1994). Weighted essentially non-oscillatory schemes. *Journal of Computational Physics*, 115(1), pp. 200–212.

Voss, C.I. (1984). SUTRA – Saturated-Unsaturated Transport, A finite-element simulation model for saturated-unsaturated, fluid-density-dependent groundwater flow with energy transport or chemically-reactive single species solute transport. *U.S. Geological Survey Water-Resources Investigation Report*, 84, p. 4369.

Voss, C.I. (1999). USGS SUTRA code – History, practical use, and application in Hawaii. In Bear, J., Cheng, A.H.-D., Sorek, S., Ouazar, D. and Herrera, I. (eds.), *Seawater Intrusion in Coastal Aquifers – Concepts, Methods and Practices*. Kluwer Academic Publishers, Boston, pp. 249–313.

Voss, C.I. and Provost, A.M. (2002). *A Model for Saturated-Unsaturated Variable-Density Ground-Water Flow with Solute or Energy Transport*, Water-Resources Investigations Report, No. 2002-4231, US Geological Survey, Reston, VA.

Chapter 4

Ideal reactors and simple water environment modeling

4.1 CONCEPTS AND MATHEMATICAL SIMULATION OF IDEAL REACTORS

Ideal continuous-flow reactors are constructed as chemical processing and water treatment units. With known reaction kinetics, these reactors convert influent water quality to desirable effluent quality by adjusting its hydraulics. The most common continuous-flow ideal reactors are CSTR and PFR. To be a CSTR reactor, the dispersion coefficient of the reactor is assumed to be infinitively large. To be a PFR reactor, the cross-sectional dispersion coefficient is assumed to be infinitely large, while the longitudinal dispersion coefficient of the reactor is assumed to be zero.

Natural water bodies such as rivers, lakes, reservoirs, or groundwater aquifers do not satisfy exactly the assumptions of ideal reactors. However, water environment models developed on the basis of ideal reactor concepts have been widely used because they are effective analytical tools for preliminary modeling analyses. Also, results of simple ideal reactor modeling for a water environment system provide guidance and useful references for further development and calibration of more comprehensive models.

4.1.1 Hydraulic residence time distribution and mean hydraulic residence time

The hydraulic property of an ideal reactor is represented by mean hydraulic residence time or mean hydraulic detention time, which is the time a water molecule would reside in the reactor. Mathematically, it is defined as follows:

$$t_R = \frac{V}{Q} \tag{4.1}$$

where V is the reactor volume and Q is the flowrate.

DOI: 10.1201/9781003008491-4

A real reactor behaves as a nonideal continuous-flow reactor system. For this system, entering water particles undergo different flow paths and eventually exit from the reactor at different time. Thus, the hydraulic properties of these nonideal reactors have to be evaluated in terms of the residence-time distribution function, which is known in statistics as the probability density function (Clark, 1996).

The residence-time distribution function of a nonideal continuous-flow reactor is usually determined by tracer experiments using a conservative tracer such as Rhodamine WT dye. In tracer experiments, tracer mass of a given concentration is released as a pulse input at the entrance of a reactor, and the tracer concentration at the outlet is then measured to plot the concentration versus time curve (Crittenden et al., 2005). The tracer concentration versus time curve measured by a tracer experiment can be taken as the probability density function $f(t)$ if the measurements are taken at very small time increments. The first moment of the function is the mean residence time, t_{res}:

$$t_{res} = \int_0^\infty f(t)t\,dt \qquad (4.2)$$

Comparing the mean residence time t_{res} of a real water environment system with the theoretical hydraulic residence time t_R indicates how closely the water environment system behaves as an ideal reactor.

4.1.2 Completely stirred tank reactor

Figure 4.1 is the schematic diagram of a CSTR. An impeller symbol is added to the diagram to indicate perfect mixing. Q_{in} and Q_{out} are the rates of inflow and outflow, respectively; C_{in} and C_{out} are the concentrations of a transporting substance in inflow and outflow, respectively; and V is reactor volume. Note that the concentration of the transporting substance in a CSTR only changes with respect to time, but not changes spatially. Also, the outflow substance concentration from a CSTR equals the substance concentration in the reactor or $C_{out} = C$.

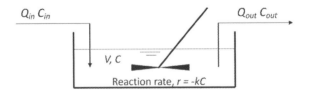

Figure 4.1 The schematic diagram of an ideal completely stirred tank reactor.

The governing equation of a CSTR reactor can be derived on the basis of mass balance principle:

$$\frac{\partial M}{\partial t} = F(t)V + rV \tag{4.3}$$

where $M = VC$ is the mass of the transporting substance in a CSTR reactor, $F(t)V$ is mass flux, and r is reaction rate, which is assumed to be first-order in this section and is expressed mathematically as: $r = -kC$.

With these definitions, Eqn. (4.3) takes the form of an ordinary differential equation:

$$\frac{d(VC)}{dt} = Q_{in}C_{in} - Q_{out}C_{out} + V(-kC) \tag{4.4}$$

Equation (4.4) has two unknown variables V and C. Therefore, it must be solved simultaneously with another independent flow equation. If the flow is steady or $Q_{in} = Q_{out} = Q$, then V is also a constant and Eqn. (4.4) becomes:

$$\frac{dC}{dt} = \frac{Q}{V}C_i - \frac{Q}{V}C - kC \tag{4.5}$$

Equation (4.5) is often expressed alternatively in the following form:

$$\frac{dC}{dt} + \left(\frac{Q}{V} + k\right)C = \frac{Q}{V}C_i \tag{4.6}$$

A CSTR model of a water environment system simulates the system's transport characteristics by relevant reaction kinetics and hydraulics. In a simple CSTR model based on Eqn. (4.6), the system's transport characteristics are represented completely by a single term of $Q/V + k$, where Q/V denotes hydraulics and k denotes the reaction kinetics. In general, the transport characteristics of a CSTR model are represented by a characteristic value λ, in this case, $\lambda = Q/V + k$.

As discussed in the last section, V/Q of a CSTR model is defined as its mean hydraulic residence time t_R. Therefore, the characteristic value can also be expressed as $\lambda = k + 1/t_R$.

With the definitions of λ and t_R, Eqn. (4.6) becomes:

$$\frac{dC}{dt} + \lambda C = \frac{C_i}{t_R} \tag{4.7}$$

Equation (4.7) is a linear ordinary differential equation. It can be readily solved by the method of integrating factor (Sokolnikoff and Redheffer, 1966).

According to the method of integrating factor, a linear ordinary differential equation in the form of Eqn. (4.7) has an integrating factor $e^{\int \lambda dt} = e^{\lambda t}$.

Multiplying both sides of Eqn. (4.7) by this integrating factor:

$$e^{\lambda t}\frac{dC}{dt} + e^{\lambda t}\lambda C = e^{\lambda t}\left(\frac{C_i}{t_R}\right) \tag{4.8}$$

Two terms on the left-hand side of the equation are merged into one term to get:

$$e^{\lambda t}\frac{dC}{dt} + e^{\lambda t}\lambda C = \frac{d}{dt}\left(e^{\lambda t}C\right) \tag{4.9}$$

Inserting Eqn. (4.9) into Eqn. (4.8), we get:

$$\frac{d}{dt}\left(e^{\lambda t}C\right) = e^{\lambda t}\left(\frac{C_i}{t_R}\right) \tag{4.10}$$

Equation (4.10) can be readily integrated with the initial condition of $C = C_0$ when $t = 0$:

$$C(t) = C_0 e^{-\lambda t} + \frac{C_i}{\lambda t_R}\left[1 - e^{-\lambda t}\right] \tag{4.11}$$

or

$$C(t) = C_0 e^{-\left(\frac{Q}{V}+k\right)t} + \frac{QC_i}{Q+kV}\left[1 - e^{-\left(\frac{Q}{V}+k\right)t}\right] \tag{4.11a}$$

where $C(t)$ is the variation of the transporting substance concentration inside the reactor; and by the definition of a CSTR, it also gives the variation of effluent concentration $C_{out}(t)$. Figure 4.2 shows the variation of the effluent concentration from a CSTR receiving constant inflow concentration C_{in}, and initial concentration in the reactor $C_0 = 0$.

As time increases, the effluent concentration approaches to a constant value, which is defined as the steady-state effluent concentration C_s. Mathematically, C_s can be determined by letting $t \to \infty$ or

$$C_s = \frac{C_{in}}{\left[1 + \dfrac{kV}{Q}\right]} = \frac{C_{in}}{1 + k(t_R)_C} \tag{4.12}$$

where $(t_R)_C$ denotes the hydraulic residence time of a CSTR reactor.

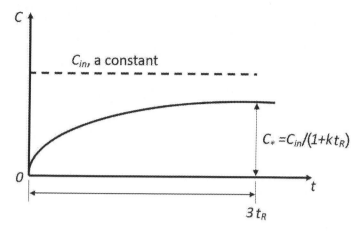

Figure 4.2 Effluent concentration (C_*) from a CSTR with first-order reaction.

In practice, the concentration at $t = 3t_R$ is used approximately as the steady-state concentration, which is about 99% of the theoretical value C_*.

4.1.3 Ideal plug flow reactor

Figure 4.3 is a schematic diagram of an ideal PFR. In this figure, Q_{in} and Q_{out} are the rates of inflow and outflow, respectively; C_{in} and C_{out} are transporting substance concentrations in inflow and outflow, respectively; A is the cross-sectional area of the reactor; l_r is the length of the reactor; and $V = Al_r$.

In a PFR, the lateral dispersion coefficient is assumed to be infinitely large and the longitudinal dispersion coefficient is assumed to be zero. As the inflow Q_{in} with a substance concentration C_{in} moves downstream, the concentration at any cross section is uniformly distributed. Therefore, a PFR is simulated by a one-dimensional model and the substance concentration inside a PFR is subject to the actions of advection and reaction (Figure 4.3).

In a PFR as shown in Figure 4.3, the concentration of a substance varies with respect to both time and distance or $C = C(x,t)$. Taking a small element ΔV in the reactor, which is formed by the cross-sectional area A and an incremental distance Δx, or $\Delta V = A\Delta x$. If Δx is very small, this small element can be considered as a completely mixed system (Figure 4.3).

The continuity equation of the flow in a small volume element in a PFR reactor takes the following form:

$$\frac{\partial(\Delta VC)}{\partial t} = QC_x - QC_{x+\Delta x} + \Delta Vr \tag{4.13}$$

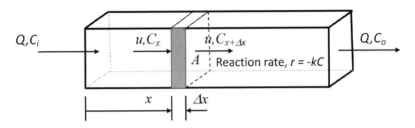

Figure 4.3 The schematic diagram of a plug flow reactor.

Divide Eqn. (4.13) by ΔV and note that $\Delta V = A\Delta x$, Eqn. (4.13) becomes:

$$\frac{\partial C}{\partial t} = \frac{Q}{A}\lim_{\Delta x \to 0}\frac{C_x - C_{x+\Delta x}}{\Delta x} + r \tag{4.13a}$$

As Δx approaches to 0, Eqn. (4.13a) becomes:

$$\frac{\partial C}{\partial t} = -u\frac{\partial C}{\partial x} + r \tag{4.14}$$

Equation (4.14) is the governing equation of a time-variable PFR model with a constant flow. Assume that the system is under the steady-state condition or $\dfrac{\partial C}{\partial t} = 0$, and that the reaction can be represented by a single first-order reaction or $r = -kC$. Then, Eqn. (4.14) can be simplified to:

$$u\frac{dC}{dx} = -kC \tag{4.15}$$

As shown in Figure 4.4, the boundary condition is $C = C_{in}$ at $x = 0$. Integrating Eqn. (4.5), we get:

$$C(x) = C_{in}e^{-k\frac{x}{u}} \tag{4.16}$$

At the outlet or $x = l_r$, $C(x) = C(l_r) = C_*$ and $\dfrac{x}{u} = \dfrac{l_r}{u} = \dfrac{l_r A}{uA} = \dfrac{V}{Q} = t_R$. Thus, Eqn. (4.16) can be expressed as follows:

$$C_* = C_{in}\exp\left[-k(t_R)_p\right] \tag{4.17}$$

where $(t_R)_p$ denotes the hydraulic residence time of a PFR reactor.

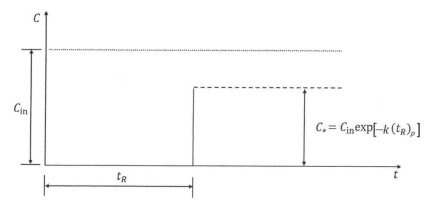

Figure 4.4 Effluent concentration (C_{out}) from a steady-state PFR with first-order reaction.

4.2   LABORATORY DETERMINATION OF REACTION KINETICS AND RATE CONSTANTS

Chemical or biochemical reaction of a transporting substance in water environment is generally categorized as a zero-order reaction, a first-order reaction, or a second-order reaction. The order of reaction and the reaction rate constant (or reaction coefficient) of a particular reaction can be determined by laboratory experiments in a batch reactor.

The order of a chemical reaction in water environment is defined mathematically as follows:

$$r = \frac{dC}{dt} = -kC^a \tag{4.18}$$

where a denotes the order of reaction. As shown later, the units of the reaction coefficients depend on the actual order of a chemical reaction.

A batch reactor is a special form of CSTR. It is formed after the inflow and the outflow of a CSTR are cut off (Figure 4.5). A CSTR, the simplest form of ideal reactors, simulates the actions of both reaction kinetics and hydraulics on the change of transporting substance in the reactor. By cutting off inflow and outflow, a batch reactor simulates only the action of reaction kinetics. In a batch experiment, the water sample with substance to be measured is placed in a batch reactor to form an initial substance concentration, and then the change of substance concentration with time is observed and measured.

In environmental engineering, enzyme-catalyzed biochemical reactions are often simulated by a more complicated saturation-type reaction, which has two reaction coefficients. Its definition and application will be discussed in Section 4.2.4.

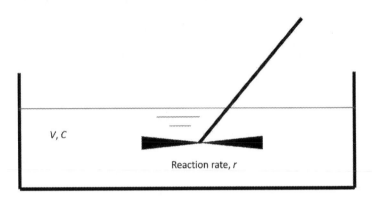

Figure 4.5 A batch reactor (CSTR).

4.2.1   Zero-order reaction

Hydrogen and chlorine combined to form hydrochloric acid in the presence of sunlight, which is one example of a zero-order reaction. A zero-order reaction can be expressed by Eqn. (4.18), with constant $a = 0$, or

$$\frac{dC}{dt} = -kC^0 = -k \tag{4.19}$$

Equation (4.19) indicates that for a zero-order reaction, the rate of reaction does not change with substance concentration. If the initial concentration is C_0, the mathematical solution of Eqn. (4.19) is expressed as follows:

$$C(t) = C_0 - kt \tag{4.20}$$

According to Eqn. (4.20), substance concentration C and reaction time t of a zero-order reaction have a straight-line relationship.

Example 4.1 Zero-order reaction and the laboratory determination of its reaction coefficient.

Problem: Experimental results of a batch experiment in terms of the relationship of substance concentration C and reaction time t are plotted in Figure 4.6 as black dots to identify the order of reaction and to determine the reaction coefficient of this reacting substance in the reactor.

Solution: Because the experimental results can be fitted satisfactorily by a straight line (Figure 4.6), the reaction can be taken as a zero-order reaction. According to Eqn. (4.20), the slope of this straight line is the negative of the reaction coefficient or $-k$ ($g/m^3/d$).

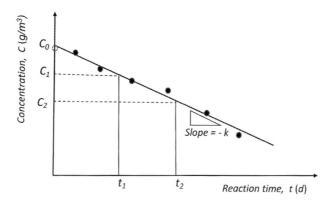

Figure 4.6 Zero-order reaction as identified by a batch experiment (Example 4.1).

Comments

1. The data points observed in a batch experiment in a graph of substance concentration C versus reaction time t are plotted. If the plot gives a straight line, the reaction is zero-order reaction, and the slope of the straight line is –k.
2. The reaction is not a zero-order reaction if experimental results do not fit well as a straight line. In this case, further analysis of experimental data must be carried out to see if the reaction can be better represented by another kinetic formulation.
3. The value of the reaction coefficient as determined by a batch experiment as shown in this example is referred to as the experimental value. Field water quality surveys are conducted to determine the field value of a reaction coefficient.

4.2.2   First-order reaction

Many natural reactions taken place in water environment such as decomposition of organic waste, atmospheric reaeration, and the coliform bacteria die off are all recognized as first-order reactions. A first-order reaction can be formulated by making the constant a in Eqn. (4.18) to be 1 or,

$$\frac{dC}{dt} = -kC^1 = -kC \tag{4.21}$$

Equation (4.21) indicates that, for a first-order reaction, the rate of reaction changes with substance concentration. If the initial concentration is C_0, the solution of Eqn. (4.20) is expressed as follows:

$$C(t) = C_0 e^{-kt} \tag{4.22}$$

According to Eqn. (4.22), the concentration of a reactant undergoing a first-order reaction decays (or grows) exponentially. By taking a logarithmic transformation, Eqn. (4.22) can be expressed as follows:

$$\ln C(t) = \ln C_0 - kt \tag{4.22a}$$

Equation (4.22a) indicates that for a substance undergoing a first-order reaction, its concentration C in the logarithmic scale or ln C and reaction time t have a straight-line relationship.

Example 4.2 First-order reaction as identified by an experimental study.

Problem: A batch experiment was conducted to determine the reaction kinetics of a reacting substance. Experimental results in terms of the relationship of substance concentration C and the reaction time t do not show a straight-line relationship. Therefore, the reaction is not a zero-order reaction. Further analysis indicates that the results can be fitted satisfactorily by a straight line in a natural logarithm of C versus t plot (see Figure 4.7).

Solution: Because the experimental results can be fitted satisfactorily by a straight line in a natural logarithm of C versus t plot (Figure 4.6), the reaction of this contaminant can be taken as a first-order reaction. According to Eqn. (4.22a), the slope of this straight line is the negative of the first-order reaction coefficient or $-k$.

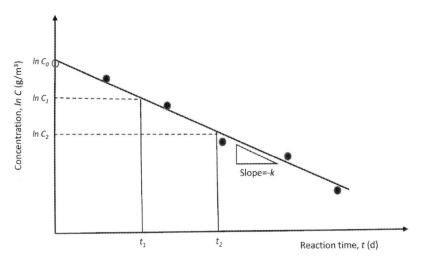

Figure 4.7 First-order reaction as identified by a batch experiment (Example 4.2).

Comments

1. The data points observed in a batch experiment in a graph of substance concentration ln C versus reaction time t are plotted. If the plot gives a straight line, the reaction is first order and the slope of the straight line is $-k$.

2. The reaction is not a zero-order reaction or first-order reaction if experimental results of a contaminant do not fit well with a straight line in either C versus t plot or ln C versus t plot. In this case, further analysis should be taken to see if the reaction is a second-order reaction.

4.2.3 Second-order reaction

Ammonium cyanate in water isomerizes into urea is one example of the second-order reaction. A second-order reaction can be formulated by making the constant a in Eqn. (4.18) to be 2 or:

$$\frac{dC}{dt} = -kC^2 \tag{4.23}$$

Equation (4.23) indicates that for a second-order reaction, the reaction rate changes with the square of substance concentration C. If the initial concentration is C_0, the solution of Eqn. (4.23) is expressed as follows:

$$\frac{1}{C(t)} = \frac{1}{C_0} + kt \tag{4.24}$$

According to Eqn. (4.24), for a second-order reaction, the inverse of substance concentration or 1/C and reaction time t have a straight-line relationship.

Example 4.3 Second-order reaction as identified by an experimental study

Problem: Results of a batch experiment of a substance in water were analyzed and indicated that they could not be fitted as a zero-order reaction or a first-order reaction. Determine the possible reaction kinetics.

Solution: Experimental results in terms of the relationship of the inverse of substance concentration 1/C and reaction time t are plotted in the Figure 4.8 as black dots. Because the experimental results can be fitted very well by a straight line, the reaction can be taken as a second-order reaction. According to Eqn. (4.24), the slope of this straight line is the reaction coefficient or k (g/m³/d).

Comments

The data points observed in a batch experiment in a graph of substance concentration ln 1/C versus reaction time t are plotted. If the plot gives a straight line, the reaction is second-order and the slope of the straight line is k.

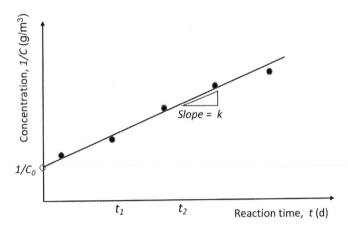

Figure 4.8 Second-order reaction as identified by a batch experiment (Example 4.3).

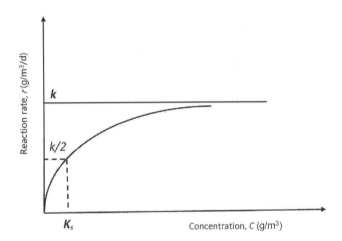

Figure 4.9 Saturation-type reaction with two reaction coefficients.

4.2.4   Saturation-type reactions

Enzyme-catalyzed biochemical reactions are often simulated by saturation-type reaction. Saturation-type reaction can be expressed mathematically by the following equation:

$$r = \frac{dC}{dt} = \frac{kC}{K_s + C} \qquad (4.25)$$

where coefficient K_s is the saturation constant (g/m³; see Figure 4.9).

If the substance (or reactant) concentration of a saturation-type reaction is very small, or the concentration C is much smaller than the saturation concentration K_s, $K_s + C \rightarrow K_s$, and Eqn. (4.25) can be simplified as follows:

$$\frac{dC}{dt} \approx (\frac{k}{K_s})C \tag{4.26}$$

Because both k and K_s are constants, k/K_s is also a constant. Therefore, if the reactant concentration of a saturation-type reaction is very small, it can be taken approximately as a first-order reaction with a single-reaction coefficient.

Conversely, if the reactant concentration of a saturation-type reaction is very large, or the concentration C is much smaller than the saturation concentration K_s, $K_s + C \rightarrow C$, and Eqn. (4.25) can be taken approximately as a zero-order reaction.

$$\frac{dC}{dt} \approx k \tag{4.27}$$

Example 4.4 Determination of the reaction coefficients of a saturation-type reaction.

Problem: An experiment was conducted in a batch reactor to study an enzyme-catalyzed biochemical reaction. Experimental results are presented:

Reactant concentration, C (mg/L)	30	15	10	7	6
Reaction rate, r (mg/L/d)	16	12	10	8	7

Assuming this reaction can be simulated as a saturation-type reaction, the values of reaction coefficients k and K_s are determined.

Solution: Rearranging Eqn. (4.25), the following algebraic equation of a straight line in terms of two variables $(1/r)$ versus $(1/C)$ can be formed:

$$\frac{1}{r} = \frac{1}{k} + \frac{K_s}{k}\frac{1}{C} \tag{4.28}$$

Experimental results can be expressed in $(1/r)$ and $(1/C)$ as presented:

1/r	0.143	0.063	0.083	0.100	0.125
1/C	0.167	0.033	0.067	0.100	0.143

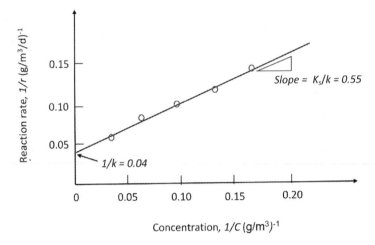

Figure 4.10 Saturation-type reaction as identified by a batch experiment (Example 4.4).

Experimental data points can be fitted well by a straight line of $(1/r)$ versus $(1/C)$ as shown in Figure 4.10. The intercept of this straight line is 0.04 and the slope is 0.55. According to Eqn. (4.28), the intercept of the straight line is $1/k$ and the slope is K_s/k.

With $1/k = 0.04$, the maximum speed of the reaction $k = 25$ mg/L/d, and with $K_s/k = 0.55$, the saturation concentration constant $K_s = 13.75$ mg/L.

Comments

1. If experimental results of a substance do not fit well as a straight line in C versus t plot, $\ln C$ versus t plot, or $1/C$ versus t plot, further analysis can be taken to see if the reaction is a saturation-type reaction.

2. Saturation-type reaction, which is often used to formulate enzyme-catalyzed biochemical reactions, contains two reaction coefficients. If the reactant concentration is very small relative to saturation concentration, a saturation-type reaction can be approximated as a first-order reaction. In a past modeling study of the environment fate and transport of dibromochloropropane (DBCP) in topsoils, the DBCP concentration in the soils was less than 100 ppt (μg/L). Successful simulation was accomplished by taking the DBCP decomposition reaction in soils as a pseudo first-order reaction (Lin et al., 1995).

4.3   FORMULATION AND APPLICATION OF SIMPLE WATER ENVIRONMENT MODELS

4.3.1   Detention ponds as CSTRs with time-variable inflow

As one of the most popular nonpoint source pollution control measures, a storm runoff detention pond provides temporal storage of storm runoff produced over a watershed to manage its flow and water quality.

By taking the concentration of the transporting chemical $C = M/V$ in a detention pond, the transport equation or Eqn. (4.4) is modified to get:

$$\frac{dM}{dt} = Q_{in}C_{in} - Q_{out}\frac{M}{V} - kM \tag{4.29}$$

In applied hydrology, storm runoff produced over a watershed is estimated and expressed by flood hydrograph analysis (see Section 5.2). Because the CSTR model of a detention pond uses flood hydrograph as its inflow, both Q_{in} and Q_{out} in Eqn. (4.29) are time-variable functions. Generally, Eqn. (4.29) contains two unknown variables of mass M and flow Q_{out}; therefore, an independent flow equation must be solved together. Based on mass conservation principle, the flow equation is formulated as follows:

$$\frac{dV}{dt} = Q_{in}(t) - Q_{out}(t) \tag{4.30}$$

As shown in Figure 4.11, volume increment dV can be expressed in terms of water depth increment dh or $dV = A(h)\,dh$. Note that $A(h)$ does not change with time and dh can be used to replace dV as an independent variable in Eqn. (4.30) to get:

$$A(h)\frac{dh}{dt} = Q_{in}(t) - Q_{out}(t) \tag{4.31}$$

Example 4.5 Modeling a detention pond system as a CSTR with time-variable flow.

Problem: A detention pond simulated as a CSTR receives inflow of watershed runoff (Figure 4.11). The runoff contains a contaminant at a concentration $C_{in} = 200$ g/m³. The contaminant undergoes a first-order decay in the pond with a reaction coefficient of $k = 0.1$ d⁻¹. Initially, the contaminant concentration in the pond is zero. The surface area and water depth of the pond is related by an empirical formulation of $A(h) = 400h^{0.7}$. Water flows out of the detention pond through a diameter culvert $d_c = 8$ in. Outflow

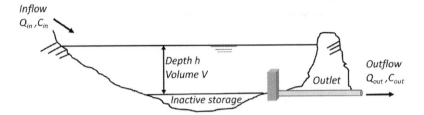

Figure 4.11 Flow and transport simulation of a storm runoff detention pond.

rate Q_{out} can be determined by an orifice formula $Q_0 = C_d A_0 \sqrt{2gh}$ in which the orifice coefficient C_d has a **value** of 0.9, and $A_0 = C_d \pi d_c^2/4$.

1. Calculate the outflow hydrograph $Q_{out}(t)$.
2. Calculate the outflow contaminant concentration over time or polluto-graph $C_{out}(t)$.

Solution: 1. Calculate outflow hydrograph, $Q_{out}(t)$

Computation in this example is carried out in an EXCEL spreadsheet as shown in Table 4.1.

With the given data, $C_d A_0$ is calculated to be 0.02918, and $2g = 2(9.8) = 19.6$. Inserting these data into the orifice formula $Q_0 = C_d A_0 \sqrt{2gh}$, we get:

$$\frac{dh}{dt} = \frac{Q_i(t) - C_0 A_0 \sqrt{2gh}}{A(h)} = \frac{Q_i(t) - 0.029\sqrt{19.6h}}{400h^{0.7}} \tag{4.32}$$

By Euler approximation, $\dfrac{dh}{dt}$ can be expressed as follows:

$$\frac{dh}{dt} \approx \frac{h(t + \Delta t) - h(t)}{\Delta t} \tag{4.33}$$

Combining Eqns. (4.32) and (4.33),

$$h(t + \Delta t) = h(t) + \Delta t \left[\frac{Q_{in}(t) - C_0 A_0 \sqrt{2gh}}{A(h)} \right] \tag{4.34}$$

Successive values of $h(t)$ at quarter-hour intervals can be calculated as illustrated in Table 4.1, starting with the initial value of depth or $h(0) = 0.5$ m as shown in column 3/row 1.

Table 4.1 Computation of the outflow hydrograph in example 4.5

(1) Time (h)	(2) Q_{in} (m³/s)	(3) h(t) (m)	(4) Q_{out} (m³/s)	(5) A-surf (m²)	(6) dh/dt (m/s)	(7) h(t + dt) (m)
0	0	0.500	0.091	246.23	−0.0004	0.166
0.25	0.18	0.166	0.053	113.85	0.0011	1.173
0.5	0.36	1.173	0.140	447.22	0.0005	1.616
0.75	0.54	1.616	0.164	559.65	0.0007	2.220
1	0.72	2.220	0.192	699.06	0.0008	2.899
1.25	0.9	2.899	0.220	842.67	0.0008	3.626
1.5	1.1	3.626	0.246	985.41	0.0009	4.406
1.75	0.99	4.406	0.271	1129.42	0.0006	4.978
2	0.88	4.978	0.288	1230.32	0.0005	5.411
2.25	0.77	5.411	0.301	1304.27	0.0004	5.735
2.5	0.66	5.735	0.309	1358.45	0.0003	5.967
2.75	0.55	5.967	0.316	1396.73	0.0002	6.119
3	0.44	6.119	0.320	1421.39	0.0001	6.195
3.25	0.33	6.195	0.322	1433.77	0.0000	6.200
3.5	0.22	6.200	0.322	1434.63	−0.0001	6.136
3.75	0.11	6.136	0.320	1424.28	−0.0001	6.004
4	0	6.004	0.317	1402.65	−0.0002	5.801
4.25	0	5.801	0.311	1369.26	−0.0002	5.596
4.5	0	5.596	0.306	1335.29	−0.0002	5.390
4.75	0	5.390	0.300	1300.69	−0.0002	5.183
5	0	5.183	0.294	1265.43	−0.0002	4.973
5.25	0	4.973	0.288	1229.46	−0.0002	4.762
5.5	0	4.762	0.282	1192.73	−0.0002	4.550
5.75	0	4.550	0.276	1155.18	−0.0002	4.335
6	0	4.335	0.269	1116.75	−0.0002	4.118
6.25	0	4.118	0.262	1077.36	−0.0002	3.899
6.5	0	3.899	0.255	1036.93	−0.0002	3.678
6.75	0	3.678	0.248	995.35	−0.0002	3.454
7	0	3.454	0.240	952.51	−0.0003	3.227
7.25	0	3.227	0.232	908.28	−0.0003	2.997
7.5	0	2.997	0.224	862.47	−0.0003	2.764
7.75	0	2.764	0.215	814.89	−0.0003	2.526
8	0	2.526	0.205	765.28	−0.0003	2.285
8.25	0	2.285	0.195	713.31	−0.0003	2.039
8.5	0	2.039	0.184	658.55	−0.0003	1.787

Column headings are described as follows:

1. Time sequence
2. Inflow data
3. Head above outlet bottom-line, in meter
4. Outflow calculated using orifice equation
5. Surface area of pond calculated using power relationship
6. dh/dt = (inflow – outflow)/(surface area)
7. $h(t + dt) = h(t) + dt[f(h,t)]$

Table 4.1 is completed by repeating the procedures for the remaining rows. Figure 4.12 shows computation results of both inflow and outflow hydrographs.

1. Calculate outflow pollutograph

    Again, by Euler approximation, $\dfrac{dM}{dt}$ is expressed as follows:

$$\frac{dM}{dt} \approx \frac{M(t + \Delta t) - M(t)}{\Delta t} \tag{4.35}$$

Combining Eqn. (4.35) and Eqn. (4.29), we to get:

$$M(t + \Delta t) = M(t) - \Delta t \left[Q_{in} C_{in} - Q_{out} C(t) - k M(t) \right] \tag{4.36}$$

where outflow $Q_{Out}(t)$ has already been calculated and presented in Table 4.1 and can be considered as part of the input data.

Similar to the outflow calculation, outflow pollutograph $C(t)$ is computed in an EXCEL spreadsheet as shown in Table 4.2. Successive values of

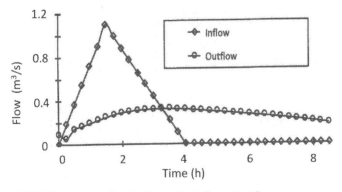

Figure 4.12 Inflow and outflow hydrographs in Example 4.5.

Table 4.2 Computation of the outflow pollutograph in example 4.5

(1) Time (h)	(2) Q_{in} (m³/s)	(3) h(t) (m)	(4) Q_{out} (m³/s)	(5) A-surf (m²)	(6) dh/dt (m/s)	(7) h(t + dt) (m)	(8) k (h−1)	(9) Cin (g/m³)	(10) Flux-I (kg/h)	(11) M(t) (g)	(12) C(t) (g/m³)	(13) Flux-o (kg/h)	(14) dM/dt (g/h)
0	0	0.500	0.091	246.23	−0.0004	0.166	0.1	100	0	0	0	0	0
0.25	0.18	0.166	0.053	113.85	0.0011	1.173	0.1	100	64.8	0	0	0	64800
0.5	0.36	1.173	0.140	447.22	0.0005	1.616	0.1	100	129.6	16200	52.508	26.445	101535
0.75	0.54	1.616	0.164	559.65	0.0007	2.220	0.1	100	194.4	41583.7	78.177	46.215	144027
1	0.72	2.220	0.192	699.06	0.0008	2.899	0.1	100	259.2	77590.3	84.992	58.895	192546
1.25	0.9	2.899	0.220	842.67	0.0008	3.626	0.1	100	324	125727	87.487	69.278	242149
1.5	1.1	3.626	0.246	985.41	0.0009	4.406	0.1	100	396	186264	88.632	78.486	298888
1.75	0.99	4.406	0.271	1129.42	0.0006	4.978	0.1	100	356.4	260986	89.169	87.042	243260
2	0.88	4.978	0.288	1230.32	0.0005	5.411	0.1	100	316.8	321801	89.317	92.681	191939
2.25	0.77	5.411	0.301	1304.27	0.0004	5.735	0.1	100	277.2	369786	89.071	96.361	143860
2.5	0.66	5.735	0.309	1358.45	0.0003	5.967	0.1	100	237.6	405751	88.536	98.607	98418
2.75	0.55	5.967	0.316	1396.73	0.0002	6.119	0.1	100	198	430355	87.775	99.72	55244.3
3	0.44	6.119	0.320	1421.39	0.0001	6.195	0.1	100	158.4	444166	86.823	99.879	14104.4
3.25	0.33	6.195	0.322	1433.77	0.0000	6.200	0.1	100	118.8	447692	85.689	99.186	25155.6
3.5	0.22	6.200	0.322	1434.63	−0.0001	6.136	0.1	100	79.2	441404	84.362	97.693	62632.9
3.75	0.11	6.136	0.320	1424.28	−0.0001	6.004	0.1	100	39.6	425745	82.812	95.404	98378.1
4	0	6.004	0.317	1402.65	−0.0002	5.801	0.1	100	0	401151	80.983	92.282	132397
4.25	0	5.801	0.311	1369.26	−0.0002	5.596	0.1	100	0	368052	78.778	88.237	125043

(1) Time (h)	(2) Q_{in} (m³/s)	(3) h(t) (m)	(4) Q_{out} (m³/s)	(5) A-surf (m²)	(6) dh/dt (m/s)	(7) h(t + dt) (m)	(8) k (h−1)	(9) C_{in} (g/m³)	(10) Flux-I (kg/h)	(11) M(t) (g)	(12) C(t) (g/m³)	(13) Flux-o (kg/h)	(14) dM/dt (g/h)
4.5	0	5.596	0.306	1335.29	−0.0002	5.390	0.1	100	0	336791	76.622	84.297	117976
4.75	0	5.390	0.300	1300.69	−0.0002	5.183	0.1	100	0	307297	74.515	80.455	111185
5	0	5.183	0.294	1265.43	−0.0002	4.973	0.1	100	0	279501	72.453	76.708	104658
5.25	0	4.973	0.288	1229.46	−0.0002	4.762	0.1	100	0	253336	70.434	73.051	98384.3
5.5	0	4.762	0.282	1192.73	−0.0002	4.550	0.1	100	0	228740	68.457	69.479	92352.6
5.75	0	4.550	0.276	1155.18	−0.0002	4.335	0.1	100	0	205652	66.519	65.987	86551.9
6	0	4.335	0.269	1116.75	−0.0002	4.118	0.1	100	0	184014	64.618	62.57	80971.2
6.25	0	4.118	0.262	1077.36	−0.0002	3.899	0.1	100	0	163771	62.75	59.222	75599.4
6.5	0	3.899	0.255	1036.93	−0.0002	3.678	0.1	100	0	144871	60.912	55.938	70425.4
6.75	0	3.678	0.248	995.35	−0.0002	3.454	0.1	100	0	127265	59.1	52.711	65437.5
7	0	3.454	0.240	952.51	−0.0003	3.227	0.1	100	0	110906	57.31	49.533	60623.8
7.25	0	3.227	0.232	908.28	−0.0003	2.997	0.1	100	0	95749.8	55.536	46.397	55971.5
7.5	0	2.997	0.224	862.47	−0.0003	2.764	0.1	100	0	81756.9	53.77	43.291	51466.8
7.75	0	2.764	0.215	814.89	−0.0003	2.526	0.1	100	0	68890.2	52.002	40.205	47094.1
8	0	2.526	0.205	765.28	−0.0003	2.285	0.1	100	0	57116.7	50.22	37.124	42835.4
8.25	0	2.285	0.195	713.31	−0.0003	2.039	0.1	100	0	46407.9	48.404	34.028	38668.8
8.5	0	2.039	0.184	658.55	−0.0003	1.787	0.1	100	0	36740.6	46.524	30.893	34566.7

$M(t) = QC_0$ at quarter-hour intervals are calculated, starting with the initial value of depth or $M(0) = 0$ as shown in column 11, row 1. Table 4.2 is completed by repeating the procedures for the remaining rows. Figure 4.13 shows both inflow and outflow pollutographs.

Comments

1. It is shown in Figure 4.12 that the inflow rates are significantly attenuated by the detention structure. It is noted that the drop in the outflow at time step 1 occurs because there was no inflow at $t = 0$, and there was already a positive head on the outlet of 0.5 m at the start of the inflow. Accordingly, the system was draining at that time.

   In the design mode, it is easy to see that for selected inflow sequences and pond and outlet dimensions, a determination can be made of flow attenuation, and the aforementioned parameters can be easily modified if the amount of attenuation is unacceptable.

2. A detention pond reduces the flood peak and the contaminant load during heavy rainstorms (Figures 4.12 and 4.13). This has been recognized as one of the most popular management measures for flood mitigation and nonpoint source pollution control.

3. Mass pollutant loading, an important parameter of nonpoint source pollution control, can be calculated by superimposing outflow hydrograph and pollutograph.

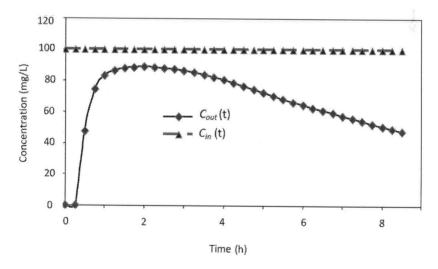

Figure 4.13 inflow and outflow pollutographs.

4.3.2   Lake water quality systems as CSTRs with steady-state flow

Lake models need to address three-dimensional variation of flow and transport and must be solved numerically (Lung, 2001; Young, 2002; Kuo et al., 2013). Popular numerical lake models include WASP (Water Quality Analysis Simulation Program) developed by the USEPA (Ambrose et al., 1993) and CE-QUAL-W2 developed by US Army Corps of Engineers (Cole, 1994). In actual lake water quality management, the response time of a lake's water quality to nutrient loading is much longer than its mean residence time. Therefore, lake systems can be often assumed as well-mixed and simple lake models based on CSTR, which have been widely used as a preliminary management tool (Vollenweider, 1975).

Figure 4.14 is a schematic diagram of a simple CSTR lake model. For simplicity, in the following discussion, it is assumed that the chemical undergoes a single first-order decomposition, or $r = -kC$.

In a simple CSTR lake system, transport characteristics are expressed by its characteristic value $\lambda = k + Q/V$, where k represents its reaction mechanism and Q/V represents its hydraulic mechanism. As discussed in Section 4.1.1, V/Q of a CSTR model is defined as its mean hydraulic residence time. Therefore, Q/V can be expressed as $1/t_R$. Thus, the characteristic value can also be expressed as $\lambda = k + 1/t_R$.

With these definitions, the governing equation of a simple CSTR lake model takes the form of Eqn. (4.7) as introduced in Section 4.1.1:

$$\frac{dC}{dt} + \lambda C = \frac{C_i}{t_R}$$

With the initial condition $C = C_0$ at $t = 0$, the aforementioned equation can be solved by the method of integrating factor to be Eqn. (4.11):

$$C(t) = C_0 e^{-\lambda t} + \frac{C_i}{1 + \lambda t_R}\left[1 - e^{-\lambda t}\right]$$

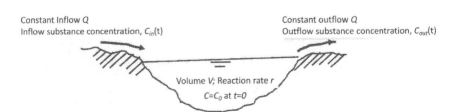

Constant Inflow Q
Inflow substance concentration, $C_{in}(t)$

Constant outflow Q
Outflow substance concentration, $C_{out}(t)$

Volume V; Reaction rate r
$C = C_0$ at $t = 0$

Figure 4.14 A simple lake substance transport system.

The simple lake system as shown in Figure 4.14 can also be simulated alternatively by following a linear system approach (see Section 2.5.3.). Figure 4.15 shows the schematic drawing of a simple linear system lake model. The model is used to simulate the variation of a transporting chemical in the lake $C(t)$ or in the outflow $C_{out}(t)$. Note that, for a CSTR, $C(t) = C_{out}(t)$.

The governing equation of a linear system lake model takes the following form:

$$C_{out}(t) = X_0(t) + \int_0^t h_1(t - \tau)X(\tau)d\tau \tag{4.37}$$

The first term on the left-hand side of Eqn. (4.37) is the zero-input response $X_0(t)$, which is the state of a lake system receiving no waste input (Figure 4.16). Generally, the initial concentration of a transporting chemical in a lake diminishes continuously by lake's self-purification ability, or the actions of flow dilution and by a first-order decomposition reaction. Therefore, the zero-input response of a simple lake system can be expressed as follows:

$$X_0(t) = C_0 e^{-\left(\frac{Q}{V}+k\right)t} = C_0 e^{-\lambda t} \tag{4.38}$$

where λ is the characteristic values of a lake system.

The second part on the left-hand side of Eqn. (4.37) is zero-state response, which is produced by system inputs and can be calculated by a convolution integration of the system input function $X(\tau)$ and the system impulse response function $h(t)$ (Figure 4.16). For a simple lake system, the loading function is expressed as $X(\tau) = QC_i/V = C_i/t_R$.

As discussed in Section 2.5, the impulse response function of a linear system can be determined by using one of the following three methods: the method of physical parameterization, the inverse method, and the method of system parameterization. Based on the method of physical parameterization, the impulse response function can be derived as the solution of the physically based model with a Dirac delta function input.

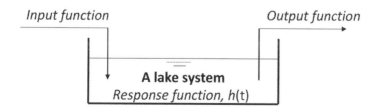

Figure 4.15 A linear systems lake transport model.

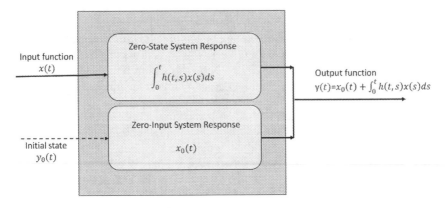

Figure 4.16 The operation of a linear system lake transporting model.

Referring to Eqn. (4.7), a simple physically based lake model with a Dirac delta function input takes the following form:

$$\frac{dC}{dt} + \lambda C = \delta(t) \tag{4.39}$$

Taking Laplace transform on both sides of Eqn. (4.39), we get:

$$sL(C) - C(0) + \lambda L(C) = 1 \tag{4.40}$$

Note that initial-state response is evaluated while the initial concentration is zero, or $C(0) = 0$, Eqn. (4.40) can be rearranged to get:

$$L(C) = \frac{1}{s + \lambda} \tag{4.41}$$

Taking a Laplace inversion, the solution back into the time domain is expressed as follows,

$$C(t) = e^{-\lambda t} \tag{4.42}$$

As the impulse response function of a linear system is the solution of the governing equation when the input function is a delta function, Eqn. (4.42) is the impulse response function or:

$$h(t) = e^{-\lambda t} \tag{4.43}$$

With the loading function of $x(t) = C_i/t_R$ and the impulse response function of $h(t) = e^{-\lambda t}$, the zero-state response of a simple lake system is express as follows:

$$\int_0^t h(t-\tau)x(\tau)d\tau = \int_0^t \left[e^{-\lambda(t-\tau)}\right]\left(\frac{C_i}{t_R}\right)d\tau = \frac{C_i}{\lambda t_R}\left[1-e^{-\lambda t}\right] \tag{4.44}$$

By superimposing the zero-input response Eqn. (4.38) and the zero-state response Eqn. (4.44), the variation of the concentration of the transporting chemical in a simple lake system, with an initial substance concentration C_0 and a constant loading $Q\,C_i/V = C_i/t_R$, is expressed as follows:

$$C(t) = C_0 e^{-\lambda t} + \frac{C_i}{\lambda t_R}\left[1-e^{-\lambda t}\right] \tag{4.45}$$

Equation (4.45) is identical with Eqn. (4.11). It indicates that the analytical solution of a simple lake system receiving a constant loading can be derived by following either a physically based approach or a linear systems approach.

Example 4.6 Lake water quality modeling.

Problem: Initially, both lakes were free of contamination. At $t = 0$, a hazardous chemical at concentration of C_{in} was found in the inflow to the first lake. If the chemical contamination in the inflow continues, calculate the outflow concentration of the chemical concentration in both lakes.

A 200,000 m^3 freshwater lake receives a constant runoff from upstream at a rate of 37,854 m^3/d. The inflow has a contaminant concentration of 200 g/m^3, which decomposes in the lake following a first-order reaction kinetics with a reaction coefficient $k = 0.1$ d^{-1}.

Calculate (1) the mean hydraulic residence time t_R, (2) the characteristic value λ, and (3) the time variation of the contaminant concentration in the lake outflow $C_{out}(t)$.

Solution: 1. The mean hydraulic residence time $t_R = \dfrac{V}{Q} = \dfrac{200,000\ m^3}{37,854\ \dfrac{m^3}{d}} = 5.28\text{d}.$

2. The characteristic value, $\lambda = \dfrac{Q}{V} + k = \dfrac{1}{5.28} + 0.1 = 0.29$ d^{-1}.

3. Apply Eqn. (4.45) and note that for an initial condition of $C_0 = 0$:

$$C(t) = \frac{C_i}{\lambda t_R}\left(1-e^{-\lambda t}\right) = \frac{200}{0.29(5.28)}\left(1-e^{-0.29t}\right)$$

$$= 130.6\left(1-e^{-0.29t}\right)\ g/m^3$$

The results are shown in Figure 4.17.

Comments

1. After a contaminant enters a lake with the inflow, the concentration in the lake increases with time, and eventually or $t \to \infty$, the concentration in the lake approaches a maximum value equals $[Q/(Q+kV)]C_i = 130.6$ g/m³.
2. In actual water quality management, the time to reach a maximum concentration is approximated as $t = 3/\lambda$. By so doing, the calculated maximum concentration would be about 95% of the theoretical value.

In water pollution control, the improvement of the water quality conditions of a polluted lake can be achieved either by reducing waste loading or by enhancing self-purification ability. One advantage of following a linear systems approach is the easiness to separate the impacts of lake's characteristics and waste loading on lake water quality variation, and therefore, it is a good management tool for water pollution control.

In this section, a constant loading function has been considered. If an exponent function is considered, the loading function takes the form of $C_i/t_R e^{-\gamma t}$. With the impulse response function remaining the same, the corresponding zero-state response of a simple lake system can be readily obtained by a small modification of Eqn. (4.44) as follows:

$$\int_0^t h(t-\tau)x(\tau)d\tau = \int_0^t \left[e^{-\lambda(t-\tau)}\right]\left(\frac{C_i}{t_R}e^{-\gamma t}\right)d\tau = \frac{C_i}{t_R(\lambda-\gamma)}\left[e^{-\gamma t} - e^{-\lambda t}\right] \quad (4.46)$$

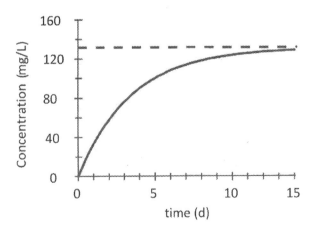

Figure 4.17 Lake water quality modeling (Example 4.6).

4.3.3 River water quality systems as PFRs with steady-state flow

According to the US Clean Water Act, every WWTP in the United States must obtain a discharge permit for its effluent. Allowable waste loads given by the permit must be determined by a modeling analysis such that the water quality standards established for the receiving river can be met, even under critical flow conditions. As the CWA administrator, the USEPA defined the critical flow of a receiving river as its minimum 7-d low flow with a 7-d recurrence interval or MA7CD10 (Liu, 1986).

In early days, WWTPs were constructed primarily to reduce organic wastes in order to maintain proper DO content in receiving rivers. DO varies along a receiving river mainly by the joint reactions of deoxygenation due to organic waste decomposition and atmospheric reaeration due to air–water interface transfer. Deoxygenation and reaeration are first-order reactions, and thus, reaction rate r can be expressed as $r = k_1L - k_2D$, where L is BOD (see Chapter 6), k_1 is deoxygenation coefficient, k_2 is reaeration coefficient, and D is DO deficit (see also Eqn. (2.92)). Because the unknown variable D varies only with downstream distance x, the PDE Eqn. (4.14) for a time-variable PFR model can be replace by an ordinary differential equation for a steady-state PFR model. By replacing the reaction rate r in Eqn. (4.14) with $r = k_1L - k_2D$, the steady-steady river BOD-DO equation takes the following form:

$$u\frac{dD(x)}{dx} = k_1L(x) - k_2D(x) \qquad (4.47)$$

Equation (4.47) indicates that the D variation in a receiving river with constant flow can be simulated by taking the river as a steady-state PFR.

Equation (4.47) contains two unknown variables L and D, and its solution requires another independent equation. Deoxygenation of BOD in a receiving river is not affected by the amount of DO; therefore, variation of BOD in a receiving river with a constant flow can also be simulated by taking the river as a steady-state PFR:

$$u\frac{dL(x)}{dx} = -k_1L(x) \qquad (4.48)$$

Equations (4.47) and (4.48) together constitute the famous Streeter–Phelps model. In practice, Eqn. (4.48) is solved by introducing proper boundary conditions to calculate in-stream variation of BOD or $L(x)$ in the receiving river, which is then introduced into Eqn. (4.47) to calculate the variation of DO in the receiving river, or $D(x)$ (refer to Chapter 6 for details).

4.3.4 Waste-assimilative capacity analysis of a water environment system as a CSTR or as a PFR

For a flow-through water environment simulated by a CSTR or a PFR, its self-purification ability is measured by the ratio of inflow and the outflow waste concentrations, or C_{in}/C_{out} as shown in Figures 4.1 and 4.3.

Equations (4.12) and (4.17) show that the outflow concentration of a transporting substance from a steady-state CSTR system and a steady-state PFR system, respectively. By rearranging these two equations, the mean hydraulic residence time $(t_R)_C$ required by a steady-state CSTR systems to convert inflow concentration C_{in} to outflow concentration C_{out} takes the following form:

$$(t_R)_C = \frac{1}{k}\left[\frac{C_{in}}{C_*} - 1\right] \tag{4.49}$$

Similarly, the mean hydraulic residence time $(t_R)_P$ required by a steady-state PFR systems to convert inflow concentration C_{in} to outflow concentration C_{out} takes the following form:

$$(t_R)_P = \frac{1}{k}\ln\left(\frac{C_{in}}{C_*}\right) \tag{4.50}$$

Because the mean hydraulic residence time of a water body is defined by Eqn. (4.1) to be $t_R = \dfrac{V}{Q}$, the efficiency of a water environment to purify entering pollution can be evaluated by the required mean hydraulic residence time to convert inflow concentration of a contaminant to a desirable outflow concentration. The smaller t_R implies a smaller water system is required to purify the contaminant and thus, is more efficient.

A diagram shown in Figure 4.18 is plotted on the basis of Eqns. (4.29) and (4.30). The abscissa in the diagram is the level of waste reduction C_* / C_{in}, and the ordinate is the ratio of the required residence time for CSTR and PFR reactors $(t_R)_C / (t_R)_P$. Figure 4.18 indicates that the ratio of required residence time for CSTR treatment unit and a PFR treatment unit to achieve 80% contaminant removal when the reaction is of first order is about 2.6. In order to achieve the same level of contaminant removal, a PFR requires much smaller residence time and thus is much more efficient. Figure 4.18 also indicates that this conclusion is valid as long as the level of removal is larger than zero.

Example 4.7 Comparison of CSTR and PFR treatment units

Problem: A water or wastewater treatment unit is built to reduce the outflow contaminant concentration under steady-state operation. A wastewater treatment unit is being built to remove 85% of a contaminant. Laboratory

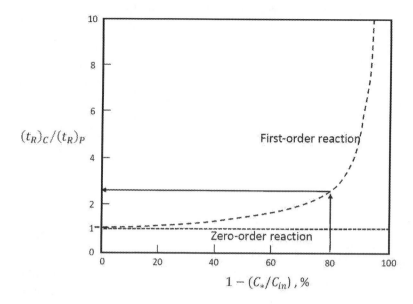

Figure 4.18 Self-purification ability of CSTR and PFR modeling systems with varying reaction rates.

analysis of this contaminant has been conducted, and it is found that it decomposes by a first-order reaction with the reaction coefficient $k = 0.1$ h^{-1}.

Calculate the required mean hydraulic residence time if the unit is designed as (1) a CSTR and (2) a PFR.

Solution: For a removal of 85% of the inflow contaminant concentration, the ratio of the waste concentrations in the outflow and the inflow is $C_{in}/C_{out} = (1.0/0.15) = 6.67$.

(1) CSTR

$$\left(t_R\right)_C = \frac{1}{k}\left(\frac{C_{in}}{C_*} - 1\right) = \frac{1}{0.1}(6.67 - 1) = 56.7\,(\text{h})$$

(2) PFR

$$\left(t_R\right)_P = \frac{1}{k}\ln\left(\frac{C_{in}}{C_*}\right) = \frac{1}{0.1}\ln(6.67) = 19\,(\text{h})$$

Comments

In order to achieve the same degree of waste reduction, a PFR requires a smaller mean hydraulic residence time than a CSTR or a smaller reactor volume under a constant flow. Thus, the PFR treatment unit provides higher treatment efficiency than a CSTR treatment unit.

EXERCISES

1. One clean lake starts receiving a steady inflow of 37,854 m³/d at $t = 0$. The inflow contains a contaminant with a concentration of 200 g/m³. The lake volume is 200,000 m³. The contaminant undergoes a fist-order decay with a reaction coefficient of 0.1 d⁻¹. A preliminary investigation indicates that the lake can be simulated as a CSTR system.

    (a) Calculate the lake characteristic value λ.
    (b) Calculate the peak outflow concentration.
    (c) Verify that at a time $t = \lambda/3$, the lake outflow concentration is 95% of the peak concentration.

2. A chemical plant discharges its waste effluent at a rate of 100 m³/d into a small well-mixed lake. The effluent contains an organic waste at a concentration of 100 g/m³, which decays in the lake as a first-order reaction with a reaction coefficient of 0.2 d⁻¹. The initial concentration of the organic waste in the lake is 20 mg/L. The lake has an outflow into a stream at a rate of 100 m³/d.

    (a) Calculate the organic waste concentration in the lake 4 d after the chemical plants start discharging its waste effluent.
    (b) Will the organic waste concentration in the lake ever reach a level of 90% of that in the plant effluent? Why?

3. A small pond has a surface area of 20000 m³, and its sides all go down vertically. Figure 4.19(a) shows the inflow hydrograph after a storm event, and Figure 4.19(b) is the relationship between the pond outflow and the water level. Calculate and plot the lake outflow hydrograph.

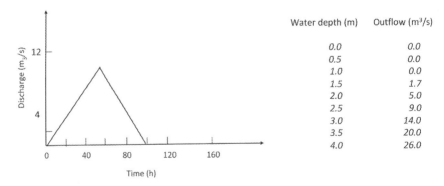

Water depth (m)	Outflow (m³/s)
0.0	0.0
0.5	0.0
1.0	0.0
1.5	1.7
2.0	5.0
2.5	9.0
3.0	14.0
3.5	20.0
4.0	26.0

Figure 4.19 Storm detention pond modeling (Exercise 3).

The initial lake water level is 1 m. Note that the lake outflow occurs only after the water is above 1 m.

4. In the previous problem, the inflow contains sediments at a concentration of 100 g/m^3. If the settling coefficient of the sediments in the lake is 1.0 m/d, calculate and plot the outflow sediments pollutograph.

5. A wastewater treatment unit is to be designed as a CSTR. The treatment objective is to reduce the contaminant concentration from 15 g/m^3 to 2 g/m^3. The contaminant reaction kinetics has been determined to be a saturation-type reaction with maximum reaction rate k = 0.02 g/m^3s, and half-saturation constant K_s = 10 g/m^3.

   (a) Calculate the required volume of the treatment unit.
   (b) Show that a saturation-type reaction can be taken as a first-order reaction when the reactant (contaminant) concentration in the unit is very low and the reaction coefficient has the form of k/K_s.
   (c) Calculate the required volume of the treatment unit if the reaction can be taken to be the first-order reaction. Comments on your results.

REFERENCES

Ambrose, R.B., Wool, T.A. and Martin, J.L. (1993). The Water quality analysis and simulation program, WASP5: Part A, model documentation. *U.S. Environmental Protection Agency*. Environmental Research Laboratory, Athens, GA.

Clark, M.M. (1996). *Transport Modeling for Environmental Engineers and Scientists*. John Wiley & Sons, New York.

Cole, T.M. (1994). *Water Operations Technical Support. Volume E-94–1. CE-QUAL-W2, Version 2.0*. Army Engineer Waterways Experiment Station Vicksburg Ms Environmental Lab.

Crittenden, J.C., Trussell, R.R., Hand, D.W., Howe, K.J. and Tchobanoglous, G. (2005). *Water Treatment – Principles and Design*. John Wiley & Sons, Inc., Hoboken, NJ.

Kuo, J., Wang, Y., Lung, W., Liu, C.C.K. and She, N. (2013). Real-time eutrophication control of a reservoir. *Proceedings of 2013 IAHR World Congress, International Association of Hydraulic Research*. Tsinghua University Press, Beijing.

Lin, P., Liu, C.C., Green, R.E. and Schneider, R. (1995). Simulation of 1, 3-dichloropropene in topsoil with pseudo first-order kinetics. *Journal of Contaminant Hydrology*, 18(4), pp. 307–317.

Liu, C.C.K. (1986). Surface water quality analysis. In Wang, L and Pereira, N.C., *Handbook of Environmental Engineering*, Vol. 4, Water Resources and Natural Control Processes. Humana Press, Clifton, NJ, pp. 1–59.

Lung, W.S. (2001). *Water Quality Modeling for Waste Load Allocation and TMDL*. Riley, New York.

Sokolnikoff, I.M. and Redheffer, R.M. (1966). *Mathematics of Physics and Modern Engineering*. McGraw-Hill, New York.

Vollenweider, R.A. (1975). Input-output models. *Schweizerische Zeitschrift für Hydrologie*, 37(1), pp. 53–84.

Young, D.L. (2002). Finite element analysis of stratified lake hydrodynamics. In Shen, H., Cheng, A.H., Wang, H.W., Teng, M., and Liu, C.C.K., *ASCE Book: Environmental Fluid Mechanics, Theories and Applications*, American Society of Civil Engineers, Reston, VA, pp. 339–376.

Chapter 5

Watershed hydrology and modeling for nonpoint source pollution control

5.1 STORM RUNOFF AND NONPOINT SOURCE POLLUTION

5.1.1 Quantity and quality of storm runoff

All the freshwater resources on the earth comes from precipitation, in the form of rain, snow, hail, and so on. For the convenience of discussion, precipitation is collectively referred to as rainfall in this book. Overland flow or runoff occurs in a watershed when rainfall exceeds infiltration. Surface runoff enters rivers and lakes, moving toward seas. In this process, if the storm runoff into a river exceeds its capacity, the excessive water will overflow to the river plain and cause floods. The primary task in designing flood prevention structures, such as earthen embankment or dam spillway, is to determine the design river discharge. Therefore, the flood hydrograph analysis, which establishes the relationship between rainfall and runoff in watersheds, is an important topic in applied hydrology.

In flood hydrograph analysis, the portion of water goes into infiltration, evaporation, and transpiration processes is called the losses, while effective rainfall is the total rainfall minus the losses. The measured rainfall of a watershed represents the average rainfall distributed evenly in the watershed. In practice, rainfall can be different in different areas of the watershed (spatial variation). Under such condition, the average rainfall of watershed is calculated by using the Thiessen polygon or other similar methods based on the gauge records of rainfall during a storm (Linsley et al., 1982). Figure 5.1(a) shows the measured rainfall in a watershed, and its two components of the losses and effective rainfall. The losses are mainly caused by infiltration. The Horton infiltration equation is often used to estimate the infiltration rate (Brutsaert, 2005).

The most widely used unit for the intensity of rainfall is mm/h. Rainfall intensity multiplied by time is the tall rainfall in mm. 1 mm of rainfall represents the total water volume in a watershed cover by 1 mm deep water, which equals the watershed area multiplied by water depth. For water resource management, all countries in the world build gaging stations to

DOI: 10.1201/9781003008491-5

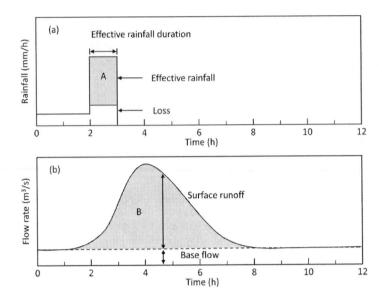

Figure 5.1 Flood hydrograph of (a) losses and effective rainfall after rainstorm and (b) baseflow and surface runoff.

measure river flow on key locations along rivers and in watersheds where surface runoff enters rivers. Figure 5.1(b) shows a flood hydrograph measured at the outlet of a watershed, consisting of two components, baseflow and surface runoff. Theoretically, the total quantity of surface runoff in the watershed is equal to the total amount of effective rainfall. Therefore, the water volume represented by 'A' in Figure 5.1(a) should be equal to the amount of water represented by 'B' in Figure 5.1(b). Thus, the quantity of storm runoff can be obtained from Figure 5.1(a) by multiplying the effective rainfall with the catchment area, which can also be obtained from Figure 5.1(b) by integrating the flow rate over time.

In the early days, the primary tasks for storm runoff management were soil conservation and flood control, focusing on water quantity. In recent years, due to the increase in water pollution, water quality of storm runoff has received more attention. The pollutants carried by runoff includes oxygen-consuming organics, suspend solids, nutrients such as nitrogen and phosphorus, *E. coil*, heavy metal, and emerging pollutants such as antibiotics and hormone. Although the concentrations of these pollutants in storm runoff are generally low, the pollution load calculated as pollutant concentration multiplied by storm runoff will have a great impact on water quality of receiving waters and form nonpoint source pollution.

Table 5.1 Quality of point source and nonpoint source pollution in urban area

Types of Waste Water	Biochemical Oxygen Demand (BOD, g/m³)	Total Soluble Solid (TSS, g/m³)	Coliform Most Probable Number (MPN/100mL)	Total Nitrogen (TN, g/m³)	Total Phosphorus (TP, g/m³)
Untreated urban sewage	200	200	5×10^7	40	10
Urban sewage after primary treatment	135	80	2×10^7	35	8
Urban sewage after secondary treatment	30	30	1×10^3	30	5
Urban sewage/rainwater combined water	115	410	5×10^6	11	4
Urban surface runoff	30	630	4×10^5	3	1

Table 5.1 presents the typical water quality parameters of point source and nonpoint source pollution discharged into the receiving water in a certain city. The data in Table 5.1 represent runoff water quality of urban catchments. As for runoff water quality in agricultural catchments, which is highly dependent on soil features and land use types, the topic will be discussed in Section 5.4.

5.1.2 Watershed modeling for nonpoint source pollution control

For the control of point source pollution such as urban sewage and industrial wastewater, the quantity and quality of wastewater discharged are the design parameters for WWTPs, which can be regarded as a pre-known constant. Moreover, the USEPA established the MA7CD10 (minimum average 7-d consecutive flow with a 10-year return period) as the design flow rate for water quality modeling of the pollution-receiving rivers. Therefore, when conducting water quality modeling, rivers are assumed to be in a steady state where river discharges are constant low flows as specified by MA7CD10. The reason for USEPA's adoption of MA7CD10 as a design flow is that the self-purification capability of a river is the lowest in dry seasons. If water quality standard can be met during dry seasons, the river water quality can be maintained at all times.

Compared to point source pollution, water quality modeling for nonpoint source pollution is more complex. The nonpoint source pollution of rivers is caused by the pollutants carried by the surface runoff during a rainstorm. Therefore, storm runoff and pollution concentration entering the rivers are largely related to the features of rainfall and catchment area. In such cases, runoff and pollution concentration cannot be regarded as known variables but need to be calculated using numerical modeling. In addition, the flow of receiving river is in a transient state after a rainstorm. The MA7CD10 can no longer be used as design flow for water quality modeling of non-point source pollution. In fact, as for water quality modeling of nonpoint source pollution, there is no widely accepted standard for the design flow condition.

Since the 1990s, the focus of water environment protection has been shifted gradually from prevention and control of point source pollution to nonpoint source pollution. For the aforementioned reasons, the numerical simulation analysis of nonpoint source pollution includes two parts, water-sheds and rivers. The purpose of the former is to estimate waste loadings entering rivers after rainstorm, which is the product of runoff and pollution concentration. The goal of river water quality modeling is to estimate the impact of the aforementioned waste loadings on river water quality. The topic of this chapter is watershed hydrology and waste loading modeling, while river water quality modeling for the control of point and nonpoint source pollution will be discussed in Chapter 6.

In order to manage nonpoint source pollution effectively, the Clean Water Act of USA established a management requirement, known as TMDL. This requirement contains two obligations for each state government: (1) to conduct water quality surveys of rivers (as well as harbors and lakes) within each state and list all water quality impaired stream segments. Water quality impaired stream segments are defined as river sections that fails to meet its water quality standard while all point source pollution have received secondary treatment; (2) to estimate TMDL for each water quality impaired stream segments, determine pollutant reduction target, and allocate load reductions to the point and nonpoint sources of pollution (Figure 5.2).

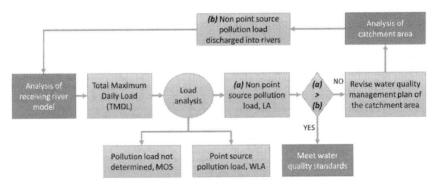

Figure 5.2 Prevention and control of nonpoint source pollution and model analysis.

A TMDL is the maximum amount of a pollutant allowed to enter a waterbody so that the waterbody will meet and continue to meet water quality standards for that pollutant. The TMDL equation is expressed as follows:

$$TMDL = WLA + LA + MOS \qquad (5.1)$$

where WLA is the sum of waste load allocations for point sources, LA is the sum of load allocations for nonpoint sources, and MOS is the margin of safety considering other uncertainties.

When TMDLs of an impaired stream segment is determined, the waste load can be allocated among different pollutant sources using procedures shown in Figure 5.2. First, watershed water quality modeling is conducted to calculate nonpoint source load entering the river (noted as LA-b). Second, river water quality modeling is performed using LA-b as input data, and TMDL and WLA are determined. Third, the TMDL equation is used to calculate LA (noted as LA-a), which is the waste loading allowed for the water watershed. If LA-b is greater than LA-a, watershed management should be strengthened to reduce waste load.

5.2 LINEAR SYSTEMS APPROACH TO WATERSHED RAINFALL–RUNOFF ANALYSIS

5.2.1 Unit hydrograph method

The IUH model is used to determine the design discharge of large hydraulic facilities. A unit hydrograph is defined as the hydrograph of direct runoff resulting from one unit depth (usually taken as 1 cm) of effective rainfall occurring uniformly over the watershed at a uniform rate for a specified duration. The unit hydrograph method was proposed by Sherman (1932), an American water conservation engineer. Sherman (1932) found that the unit hydrograph of a watershed can be calculated on the basis of the previous rainfall and discharge records of the watershed measured during a rainstorm.

There are three steps to estimate the unit hydrograph of a watershed: (1) to calculate effective rainfall from rain gauge data, (2) to compute direct surface runoff hydrograph based on discharge records, and (3) to estimate unit hydrograph using effective rainfall and runoff hydrograph. Steps 1 and 2 have been discussed in Section 5.1, and Step 3 is based on the linear system theory and the principles of superposition and proportionality.

According to the superposition principle of a linear system, if the input α leads to the output β, the input $m\alpha$ would result in the output $m\beta$. For example, if 1 cm of excess rainfall produces a direct runoff peak of 1 m³/s, 2 cm of excess rainfall would produce a direct runoff of 2 m³/s.

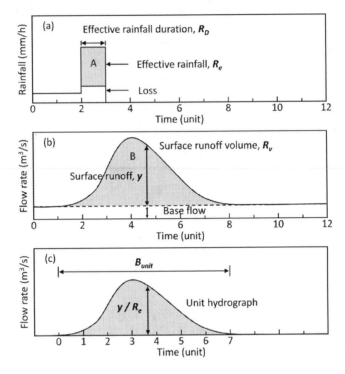

Figure 5.3 Derivation of unit hydrograph based on a set of rainfall and discharge records in a catchment area: (a) effective rainfall, (b) surface rainoff, and (c) calculation unit hydrograph.

Suppose the effective rainfall estimated in Step 1 is R_e (Figure 5.3(a)), and the direct runoff is separated from the measured hydrograph by subtracting the baseflow (Figure 5.3(b)). The unit hydrograph can be obtained by dividing the time history of direct runoff with R_e (Figure 5.3(c)). Unit hydrograph only considers the contribution of direct surface runoff; therefore, its starting point is the same as that of the effective rainfall. In practice, common units of time for a unit hydrograph include 1 h, 2 h, or 1 d. It should be noted that as the duration of the effective rainfall is R_D, theoretically the resultant unit hydrograph can be used to calculate only runoff induced by design rainfalls with the same duration R_D.

Figure 5.4 shows how to use the unit hydrograph derived earlier to calculate the runoff generated by another rainstorm in the same watershed. The duration of the effective rainfall is 3 R_D. When applying unit hydrograph,

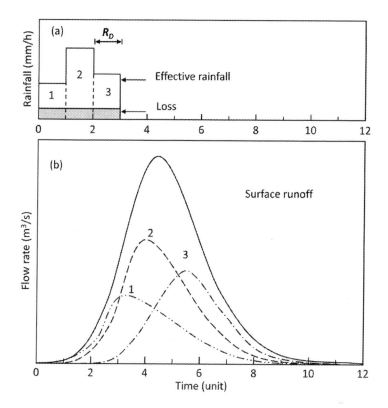

Figure 5.4 Calculation of storm discharge hydrograph by using existing unit hydrograph: (a) calculation of the effective rainfall of irregular rainstorm and (b) calculation of flow hydrograph by superposition principle.

the rainstorm can be regarded as three consecutive rainfalls with the same duration R_D. Each of the three rainfalls are regarded individually as the input of the linear watershed system, and consequently, three runoff hydrographs can be calculated using the unit hydrograph. Superposition (sum) of the three hydrographs would result in the direct runoff hydrograph of the rainstorm with varying rainfall rates (Figure 5.4(b)).

Example 5.1 The runoff hydrograph of a 3.0 square mile watershed after a rainstorm is given in Figure 5.5. Rainfall records show the watershed received an effective rainfall of 1.4 in. with a duration of one unit time.

Problem: How to obtain the unit hydrograph of this watershed?

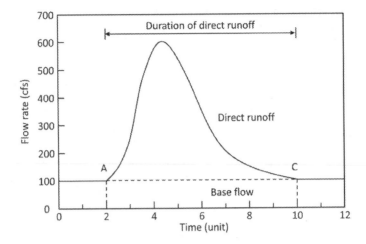

Figure 5.5 Known flow rate for Example 5.1.

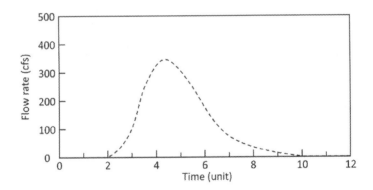

Figure 5.6 Unit hydrograph derived for Example 5.1.

Solution: (a) Draw a line A-C to separate the hydrograph into direct runoff and baseflow. As the direct surface runoff is caused by the effective rainfall, point 'A' coincides with the starting point of the effective rainfall. First, the line A-C can be assumed to be a horizontal line, and then, the location of the 'C' point is adjusted to ensure the volume of runoff and effective rainfall equal to each other.

(b) The time delay between 'A' and 'C' is the duration of direct runoff.
(c) Divide the vertical coordinate of runoff hydrograph by the effect rainfall 1.4 to get the vertical coordinate of unit hydrograph, as shown in Figure 5.6.

5.2.2 Instantaneous unit hydrograph method

The unit hydrograph method is widely used in the design of flood control facilities. However, the original unit hydrograph method proposed by Sherman (1932) is an empirical method, which has many weaknesses and limitations in practice. The biggest weakness of the unit hydrograph method is related to the duration of effective rainfall, as Huang (1937) found that the unit hydrograph of a watershed is highly dependent on the duration of effect rainfall, which affects the runoff duration. The runoff duration (B_{unit}) is dependent on the characteristics of both watershed and rainfall.

$$B_{unit} = t_c + R_D \qquad (5.2)$$

where B_{unit} is shown in Figure 5.3(c) and t_c is the time of concentration, calculated as the time for runoff to flow from the most hydraulically remote point of the watershed to the point under investigation (usually the watershed outlet).

The time of concentration reflects the characteristics of watershed, and R_D is the feature of rainfall. Different rainfall duration leads to different runoff duration. The unit hydrograph assumes that the total volume of runoff equals the total amount of effective rainfall, and consequently, when runoff duration changes, the shape of unit hydrograph will vary. Therefore, for a watershed, the unit hydrograph is not unique, and it cannot be regarded as the true impulse response function of a linear watershed system.

In order to overcome such weakness, Huang (1937) proposed the concept of IUH. After years of development in applied hydrology, the original unit hydrograph method by Sherman (1932) has been gradually revised to be the IUH method.

The IUH method assumes the watershed as a linear system, and its governing equation is as follows:

$$Y(t) = \int_0^t h(t - \tau)X(\tau)d\tau \qquad (5.3)$$

where the effective rainfall $X(\tau)$ is the input function, the direct surface runoff $Y(t)$ is the output function, and the IUH $h(t - \tau)$ is the impulse response function.

Application and superposition principle of the IUH method are illustrated in Figure 5.7. First, let's consider the incremental volume of water for the effective rainfall $X(\tau)$ over a short interval $d\tau$, termed as $[X(\tau)d\tau]$. When $d\tau$ is very small, the effective rainfall can be regarded as an instantaneous function (Figure 5.7(a)). At this specific time τ, the direct runoff generated by the incremental rainfall–volume is $X(\tau)h(t - \tau)$ (Figure 5.7(b)). Based on the superposition principle of the linear system theory, the output function runoff can be calculated using Eqn. (5.3).

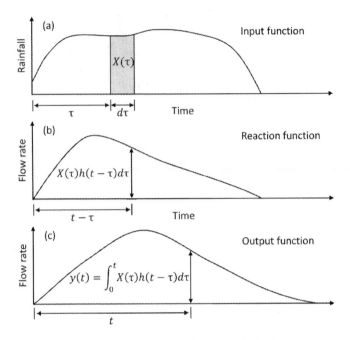

Figure 5.7 Application and superposition principle of linear hydrological system model: (a) input function, (b) it should be equal to the surface runoff of IUH in small rain–all unit, $X(\tau)h(t - \tau)$, and (c) output function.

The process of estimating the response function of a linear system, also known as system identification, is one of the most important tasks in system modeling. Three types of methods for system identification are introduced in Chapter 2, namely, the method of physical parameterization (white box), the inverse method (black box), and the method of system parameterization (gray box). In the past, the most popular methods for system identification of watershed linear systems are the inverse methods. By the inverse method, the rainfall and surface runoff data measured during a past rainstorm event is used as the input and output functions of the system, respectively. The instantaneous response function is calculated using the inverse operation of compound integral. One weakness of the instantaneous response function obtained using the inverse method is lack of robustness. Although the runoff hydrograph can be reconstructed using the resultant instantaneous response function, the result is often less than satisfactory when it is applied to predict the runoff hydrograph of another rainstorm by performing compound integration of Eqn. (5.3) with the effective rainfall.

5.2.3 Parameterization of instantaneous unit hydrograph

In order to improve the prediction ability of the reverse methods, the method of system parameterization was introduced. Nash (1959) proposed the concept of linear reservoirs to represent the flow and storage of water in the watershed, as shown in Figure 5.8. Based on this concept, the instantaneous response function for the watershed is essentially a gamma probability distribution.

Example 5.2 Nash (1959) conceptualized the relationship between rainfall and runoff in a watershed and assumed the rainfall reaching the surface flowed through a series of reservoirs before reaching the watershed outlet (Figure 5.8). The reservoirs in the process were assumed to be linear and well mixed.

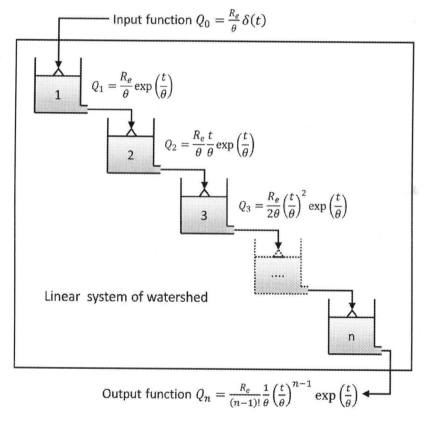

Input function $Q_0 = \frac{R_e}{\theta} \delta(t)$

$Q_1 = \frac{R_e}{\theta} \exp\left(\frac{t}{\theta}\right)$

$Q_2 = \frac{R_e}{\theta} \frac{t}{\theta} \exp\left(\frac{t}{\theta}\right)$

$Q_3 = \frac{R_e}{2\theta} \left(\frac{t}{\theta}\right)^2 \exp\left(\frac{t}{\theta}\right)$

Linear system of watershed

Output function $Q_n = \frac{R_e}{(n-1)!} \frac{1}{\theta} \left(\frac{t}{\theta}\right)^{n-1} \exp\left(\frac{t}{\theta}\right)$

Figure 5.8 The rainfall/runoff system of reservoir catchment simulated by a series of linear reservoirs.

Problem: Verify that based on the concept of linear reservoirs the IUH would take the form of a gamma distribution.

Solution: IUH is the output function of a watershed receiving one unit of instantaneous effective rainfall. Instantaneous effective rainfall can be written as $R_e \delta(t)$. If the first reservoir receives the effective rainfall of R_e, the outflow from this reservoir can be expressed as follows:

$$Q_1 = \frac{R_e}{\theta} \exp\left(\frac{t}{\theta}\right) \tag{5.4}$$

By using the law of mass conservation, the governing equation of the second reservoir can be written as follows:

$$\frac{dQ_2}{dt} + \frac{1}{\theta} Q_2 = \frac{R_e}{\theta} Q_1 \tag{5.5}$$

Substituting Eqn. (5.5) into Eqn. (5.4), we can obtain:

$$\frac{dQ_2}{dt} + \frac{1}{\theta} Q_2 = \frac{R_e}{\theta} \exp\left(\frac{t}{\theta}\right) \tag{5.6}$$

Under the initial condition of $\theta = 0$ at $t = 0$, Eqn. (5.6) can be solved by the integral factor method (refer to the discussion in Section 6.3.1 for the solution of integral factor method). The solution is expressed as follows:

$$Q_2(t) = \frac{R_e}{\theta} \frac{t}{\theta} \exp\left(\frac{t}{\theta}\right) \tag{5.7}$$

Equation (5.7) becomes the input function for the third reservoir. By using the initial condition of $\theta = 0$ at $t = 0$, the outflow of the third reservoir is expressed as follows:

$$Q_3(t) = \frac{R_e}{2\theta} \left(\frac{t}{\theta}\right)^2 \exp\left(\frac{t}{\theta}\right) \tag{5.8}$$

Based on the aforementioned calculation, the outflow function of the nth reservoir can be deduced as follows:

$$Q_n(t) = \frac{R_e}{(n-1)!} \frac{1}{\theta} \left(\frac{t}{\theta}\right)^{n-1} \exp\left(\frac{t}{\theta}\right) \tag{5.9}$$

When the number of reservoirs increases, the time history of outflow equation (5.9) becomes a gamma distribution (as shown in Figure 5.9).

Nash (1959) generalized the rainfall–runoff system in the watershed as a series of linear reservoirs. Figure 5.9 shows that when the number of reservoirs increases, the output function satisfies a gamma distribution.

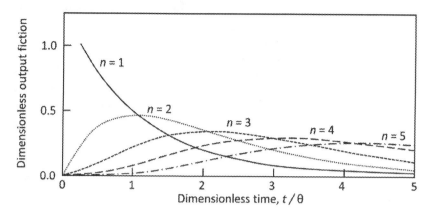

Figure 5.9 Output function of sequential linear reservoir model system.

$Q_n(t)$ in Eqn. (5.9) is the output function of the watershed (the entire linear reservoirs) after receiving the instantaneous input of rainfall R_e. According to linear system theories, if the input function is a unit of rainfall or $R_e = 1$, $Q_n(t)$ is the unit instantaneous response function of the watershed. Mathematically $(n-1)!$ is called the gamma function $\Gamma(\alpha)$. Replacing n and θ by α and β, Eqn. (5.9) can be written as follows:

$$h(t) = \frac{1}{\Gamma(\alpha)} \frac{1}{\beta} \left[\left(\frac{t}{\beta} \right)^{\alpha-1} \exp\left(-\frac{t}{\beta} \right) \right] \tag{5.10}$$

Equation (5.10) is a gamma distribution function with two parameters, also known as Nash IUH. In this equation, α is the scale parameter and β is the shape parameter (Brutsaert, 2005).

Nash's assumption is in fact system parameterization, which converts the estimation of instantaneous response function of a watershed into the estimation of two parameters α and β in the gamma distribution function.

5.3 GEOGRAPHIC INFORMATION SYSTEM AND WATERSHED MODELING

5.3.1 Geographic information system

5.3.1.1 Definition of GIS

A broadly accepted definition of geographic information system (GIS) is the one provided by the National Centre of Geographic Information and Analysis: a GIS is a system of hardware, software, and procedures to facilitate the management, manipulation, analysis, modeling, representation, and display of georeferenced data to solve complex problems regarding planning and management of resources (NCGIA, 1990). GISs have emerged in the last decade as an essential tool for urban and resource planning and management. Their capacity to store, retrieve, analyze, model, and map large areas with huge volumes of spatial data has led to an extraordinary proliferation of applications. GISs are now used for land use planning, utilities management, ecosystems modeling, landscape assessment and planning, transportation and infrastructure planning, market analysis, visual impact analysis, facilities management, tax assessment, real estate analysis, and many other applications. Functions of GIS include data entry, data display, data management, information retrieval, and analysis.

5.3.1.2 GIS applications

Mapping locations: GIS can be used to map locations. GIS allows the creation of maps through automated mapping, data capture, and surveying analysis tools.

Mapping quantities: People map quantities, like where the most and least are, to find places that meet their criteria, or to see the relationships between places. This gives an additional level of information beyond simply mapping the locations of features.

Mapping densities: While you can see concentrations by simply mapping the locations of features, in areas with many features it may be difficult to see which areas have a higher concentration than others. A density map lets you measure the number of features using a uniform areal unit, such as acres or square miles, so you can clearly see the distribution.

Finding distances: GIS can be used to find out what's occurring within a set distance of a feature.

Mapping and monitoring change: GIS can be used to map the change in an area to anticipate future conditions, decide on a course of action, or to evaluate the results of an action or policy.

5.3.2 The application of GIS in instantaneous unit hydrograph analysis

The coefficients α and β of IUH can be estimated using the observed rainfall and discharge data of a watershed (Nash, 1959). However, if rainfall and discharge records of the watershed are unavailable, some other methods are needed to synthesize the IUH.

One method to synthesize the IUH for data-scarce area is illustrated as follows: (1) select a few watersheds with abundant hydrologic observation records, and calculate the coefficients α and β of IUH for each watershed; (2) analyze hydrologic factors of each watershed that affects the generation of surface runoff, and generalize these factors into several dimensionless parameters; and (3) perform regression analysis between the α and β

coefficients and the dimensionless parameters for the selected watersheds. The resultant regression equations can be used to synthesize the IUH for data-scarce watershed (Liu et al., 2012).

The common dimensionless parameters that affect the rainfall–runoff relationship of a watershed includes: s, the average slope of the watershed; N_{SCS}, the SCS curve number, which is an indicator of the ground and soil properties; A/l_r^2, the geometric factor representing the shape of the watershed, where A is the area of the watershed, and l_r is the length of the main flow channel in the watershed.

The general form of the resultant regression equation is expressed as follows:

$$\alpha = a_0 s^{a_1} N_{scs}^{a_2} \left(\frac{A}{l_r^2}\right)^{a_3} \tag{5.11}$$

$$\beta = b_0 s^{b_1} N_{scs}^{b_2} \left(\frac{A}{l_r^2}\right)^{b_3} \tag{5.12}$$

In Eqns. (5.11) and (5.12), a_1, a_2, a_3, b_1, b_2, and b_3 are regression coefficients.

Figure 5.10 illustrates the integrated application of the IUH model and the GIS to establish a regional regression equation. First, the three dimensionless parameters, s, N_{SCS}, and A/l_r^2, that represents the characteristics

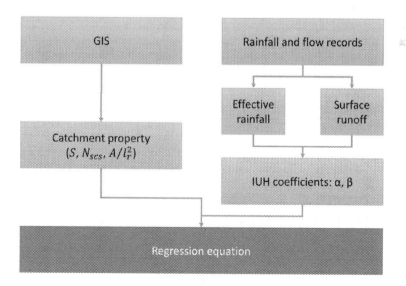

Figure 5.10 Comprehensive application of linear hydrological system model and GIS.

of watersheds are calculated using data in GIS; second, the coefficients α and β of IUH for each selected watersheds are computed using the rainfall and discharge records of the watershed; finally, the coefficients in the regression equation are determined by conducting regression analysis between the dimensionless parameters and the coefficients α and β of selected watersheds.

5.4   BASINS: A GIS-BASED WATER ENVIRONMENT MODELING PLATFORM

The USEPA developed a GIS-based water environment modeling platform, called BASINS. BASINS provides a framework that brings together modeling tools and environmental spatial and tabular data into a GIS interface. BASINS can be used for investigations and analysis on a variety of geospatial scales from small watersheds within a single municipality, to a large watershed across several states. The applications of BASINS together with other water quality models in watershed modeling are illustrated in the following sections.

5.4.1   The integration of BASINS and PLOAD model

PLOAD is one of the water quality modes incorporated in the BASINS platform. The structure of PLOAD is very simple and can be used to calculate the annual average runoff and pollution load discharged from a watersheds. It is an effective tool for preliminary numerical modeling in watershed water quality management (USEPA, 2001; Liu, 2007).

The governing equation for PLOAD is as follows:

$$L_P = \sum_n \left(R \cdot R_p \cdot f_{rn} \cdot C_n \cdot A_n \cdot \frac{2.72}{12} \right) \tag{5.13}$$

where L_P is the pollutant load (lb), R is the rainfall (in/y), R_p is the ratio of runoff to total rainfall (default = 0.9), f_{rn} is the runoff coefficient for n-type land use (inchesrun/inchesrain), C_n is the event mean concentration for the n-type land use (mg/L), and A_n is the area of n-type land use. In BASINS, areas calculated from GIS data are represented in square meters. PLOAD converts areas from square meters to acres prior to using the information in Eqn. (5.13).

The following example illustrates the application of BASIANs and PLOAD in watershed water quality modeling. The Waiawa watershed is located in the south-central part of Hawaii's Oahu Island (Honolulu City), with the lower half adjacent to Pearl Harbor and the upper half extending into the Koolau mountain range, covering a total area of approximately 25 square miles (65 km^2), as shown in Figure 5.11. The average annual rainfall of the watershed is 46 in. (1168 mm), and the resultant

Figure 5.11 Location of Waiawa catchment on Oahu Island, Hawaii.

runoff is discharged by two main streams and their tributaries. In the early stage, the lower part of the watershed was mainly used as sugarcane fields, except that a small area has been developed to be residential area. These sugarcane fields have been abandoned since 1983 and gradually replaced by shrub and grassland. The upper part of the catchment is covered by frosts.

Based on the GIS data, the Waiawa watershed can be subdivided into nine subdrainage areas (Figure 5.12(a)). The proposed land development plan was to have high-intensity development and low-intensity development in the four subdrainage areas 5, 6, 7, and 8 (Figure 5.12(b)), with a total development area of 5,100 acres or 2,060 hectares.

The BASINS-PLOAD model was implemented to calculate and display the impact of the proposed development on the annual average runoff and pollution load flowing into the Pearl Harbor. The land use types of Waiawa watershed before and after the proposed land development were mapped in BASINS (Figure 5.13).

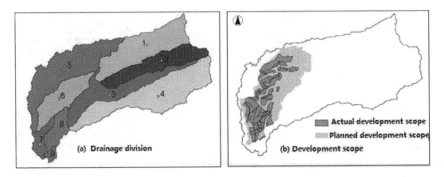

Figure 5.12 (a) Drainage division and (b) land development plan scope of the Waiawa catchment.

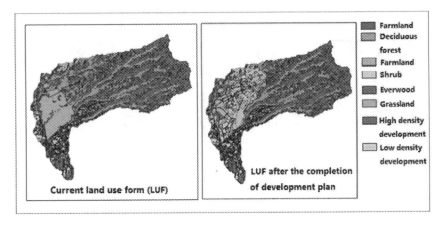

Figure 5.13 Current and planned land use patterns in the Waiawa catchment area.

In order to estimate the impact of land development on runoff and pollution load, it was necessary to compare the runoff and pollution load before and after the land development. Equation (5.11) is used to calculate the relevant pollutant load, including total suspended solids, total nitrogen, and total phosphorus. The product of R, R_p, and f_{rn} in Eqn. (5.13) was the effective runoff. The runoff coefficient f_{rn} in Eqn. (5.13) was related to the size of impervious areas in the watershed. Table 5.2 shows the typical values of impervious ground percentage and the mean pollutant concentration (EMC, C_u) of pollutants of different land use types. This table was compiled using data presented in the user's manual of PLOAD (USEPA, 2001) and other relevant literature.

Table 5.2 Average pollutant concentration and impervious area percentage of different land use type

Land Use Classification	EMCs (mg/L)			Impervious Area (%)
	TSS	TP	TN	
High-density development area	83	0.40	2.7	85
Low-density development area	85	0.30	1.7	65
Farmland	780	0.20	1.1	45
Grassland	70	0.20	1.1	40
Deciduous forest	32	0.14	1.1	35
Everwood	19.7	0.14	1.1	30
Swamp	19	0.14	1.1	30

Table 5.3 Calculation value of annual average suspended solids under current land use status

Drainage Zone	Area (acres)	Runoff (acres-ft/y)	Total Nitrogen TN (kg/y)	Average Runoff Concentration EMCs (mg/L)
1	2,643	3,321	4,451	1.100
2	1,766	2,213	2,966	1.100
3	1,924	2,643	3,978	1.235
4	3,588	4,844	7,375	1.250
5	3,032	4,113	5,736	1.145
6	1,469	2,086	2,843	1.119
7	455	755	1,339	1.457
8	662	1,118	1,965	1.443
9	142	278	604	1.781
Total	15,681	21,371	31,257	11.630

Tables 5.3 and 5.4 show the calculated total annual suspended solids load discharged into Pearl Harbor before and after development, respectively. If the land development plan was implemented, the annual runoff into Pearl Harbor would increase from the current 21,372 to 22,920 acres-ft. The total nitrogen discharged into Pearl Harbor would increase from 31,259 to 37,396 kg.

Comparison of Tables 5.3 and 5.4 illustrated the impact of changes of land-use types on the runoff and pollutant load discharged into the receiving water body. Although only total nitrogen was presented, the impacts on suspended solids and total phosphorus were almost the same (Liu, 2007).

Table 5.4 Calculation value of annual average suspended solids under land use after development

Drainage Zone	Area (acres)	Runoff (acres-ft/y)	Total Nitrogen TN (kg/y)	Average Runoff Concentration EMCs (mg/L)
1	2,643	3,319	4,448	1.100
2	1,766	2,213	2,966	1.100
3	1,924	2,646	4,002	1.242
4	3,588	4,915	7,522	1.256
5	3,032	4,745	8,065	1.395
6	1,469	2,505	4,772	1.546
7	455	919	1,959	1.750
8	662	1,366	3,044	1.829
9	142	292	618	1.739
Total	15,681	22,920	37,396	12.957

5.4.2   The integration of BASINS and HSPF model

HSPF, like PLOAD, is one of the water quality models included in BASINS. It can be used to simulate various natural mechanisms that affect the water quantity and quality of runoff in a watershed. The results of HSPF can be displayed in the form of hydrograph and pollutograph.

During simulation, HSPF divides the watershed into three components, namely, permeable surface, impermeable surface, and receiving water body. The permeable surface is termed as PRELND in HSPF and is used to simulate the flow and pollutant transport among surface runoff, shallow groundwater, and groundwater. The impervious surface is called IMPLND. Due to its imperviousness, IMPLND is only used to simulate the quantity and quality of surface runoff. The receiving water, called RCHRES, is used to simulate flow and pollutant transport in rivers and well-mixed lakes (including reservoirs).

BASINS-HSPF, which is the combination of HSPF and computer information management platform BASINS, is an ideal modeling tool for watershed water quantity and quality. The following example illustrates the application of BASINS-HSPF to estimate runoff and pollution load of Manoa-Palolo watershed on the island of Oahu, Hawaii.

Ala Wai Canal is located in the south of Honolulu city (Oahu Island), Hawaii. It was built to divert the rainstorm runoff to the Waikiki coast, a famous sightseeing and recreation area (Figure 5.14). Currently, the canal is a popular site for boating, fishing, hiking, and other activities. Improving the water quality of the Ala Wai Canal has become one of the key tasks of environment protection in Honolulu. The sub-watershed Manoa-Palolo

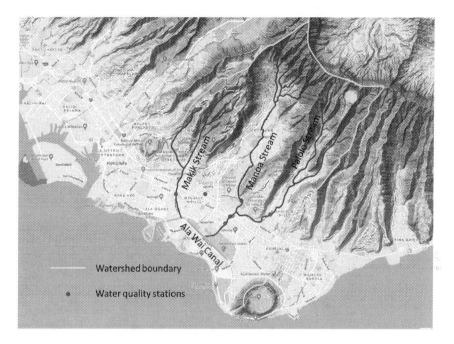

Figure 5.14 Location map of the Ala Wai Canal and its catchment area in Honolulu.

accounts for over 65% area of the Ala Mai watershed. Most of the sub-watershed is used for farming, housing, and campuses including the University of Hawaii.

The GIS data in BASINS were used to divide the Manoa-Palolo watershed into 12 drainage segments and 14 reaches (Figure 5.15). The coefficients of the model can be initially estimated using the HSPFParm database in the application manual (Donigian et al., 1999). Next, model calibration was conducted to adjust the estimated coefficients. The process of model calibration is illustrated in Figure 5.15 by using No. 7 observation station as an example (Liu et al., 2012).

The No.7 observation station was located next to a long-term US Geological Survey (USGS) gaging station. The runoff data of No.7 observation station was assumed to be same as the USGS gaging station. An ISCO automatic gauge was installed next to the USGS gaging station for water quality measurements (Figure 5.16). For other observation stations without a nearby long-term hydrologic stations, the area–velocity sensors were installed and used to measure surface runoff.

The No.7 observation station and REACH14 were in the same location in the BASINS-HSPF model (Figure 5.16). Therefore, in the model calibration

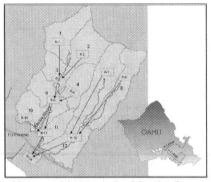
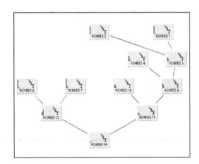

(a) Manoa-Palolo Watershed on Oahu, Hawaii   (b) BASINS-HSPF Model of Manoa-Palolo Watershed

Figure 5.15 Manoa-Palolo water collection partition and BASINS-HSPF model established by using BASINS GIS database: (a) Manoa-Palolo watershed on Oahu, Hawaii, and (b) BASINS-HSPF model of Manoa-Palolo watershed.

Figure 5.16 Location and observation equipment of the No.7 observatory in the Manoa-Palolo catchment zone.

process, the daily discharged at REACH14 calculated by BASINS-HSPF was compared to the measured discharge records at No.7 observation station, and the relevant model coefficients were adjusted to match the calculated results and measured data. Figure 5.17 shows the model calibration results using the rainstorm records of March 13–15, 2009, including runoff and total suspended solids.

The calibrated model was applied to simulate the runoff, total suspended solids, and total nitrogen pollution load flowing into Ala Wai canal in Manoa-Palolo sub-watershed, and the results are presented in Table 5.5.

The water quality model of a receiving river, which simulates the river's response to waste inputs, is developed and applied to evaluate management

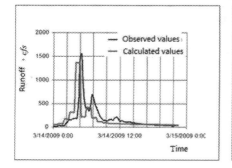
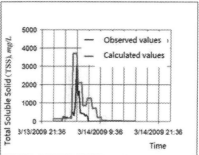

Figure 5.17 Model correction of HSPF model in Manoa-Palolo catchment partition using measured flow and water quality data after rainstorm.

Table 5.5 Runoff and pollution load into Ala Wai Canal catchment calculated by BASINS-HSPF model

Catchmen Partition Inflow Point	Partition Area (acres)	Flow (kg/d)		TSS (kg/d)		TN (kg/d)	
		Dry	Wet	Dry	Wet	Dry	Wet
Reach 1							
Honolulu Zoo	198.0	0.003	0.011	93.3	1268.1	1.9	1.9
Ala Wai Golf Course	652.0	7.960	9.204	51.6	139.4	8.9	9.8
Waikiki 1	201.0	2.400	2.792	310.6	490.3	2.0	2.0
Reach 2							
Manoa-Palolo Watershed	6809.6	10.140	19.079	4997.0	23185.4	35.8	59.0
Waikiki 2	3.0	0.002	0.009	41.0	41.0	1.1	1.1
Reach 3							
Middle	750.0	0.720	1.431	91.0	263.0	7.4	10.8
Waikiki 3	47.0	0.571	1.028	76.8	116.4	1.1	1.3
Reach 4							
Maikiki Watershed	1664.0	4.380	11.402	1526.4	4655.0	10.2	17.7
Waikiki 4	79.0	0.936	1.777	156.5	198.0	2.0	2.2

alternatives for pollution control. For point-source pollution control, modeling is generally conducted for the receiving river under critical MA7CD10 low flow conditions (refer to Section 1.2.2). Also, point source pollution is produced by municipal/industrial wastewater discharges, and their amount and quality are considered as given input parameters.

As nonpoint source pollution is caused by storm runoff, the MA7CD10 low flow cannot be used as modeling design flow. In this example, storm runoff and pollution load of the Ala Wai Canal under dry season and wet season were determined by the watershed model of BASINS-HSPF (Liu et al., 2012). The data generated by watershed modeling were then used as the input for a time-variable river model for Ala Wai Canal (see Section 9.1.1 for the model formulation).

EXERCISES

1. Select a nearby watershed, and use BASINS-PLOAD to simulate the current annual average runoff, total suspended solid, total phosphorus, and total nitrogen.

2. The following table is a 2-h unit hydrograph of a watershed:

Time (h)	Surface Runoff (m³/s)	Time (h)	Surface Runoff (m³/s)
0	0	8	116.0
2	11.3	9	90.6
3	70.8	10	67.9
4	124.5	11	45.3
5	169.8	12	25.5
6	121.3	13	11.3
7	172.6	14	5.7
8	141.5	15	0

(a) Calculate the area of the watershed.
(b) Given the effective rainfall is 3 in. with a duration of 6 h, calculate the peak flow of the runoff curve caused by this rain.

3. Assume the watershed in Example 5.1 receives a rainstorm, and its effective rainfall is shown in the following table:

Time Unit	Effective Rainfall (in)
1	0.7
2	1.7
3	1.2

If the baseflow can be neglected, please calculate and draw the runoff hydrograph caused by this rainstorm.

4. Based on Eqn. (5.7) and Figure 5.9, please calculate and prove that the maximum time of runoff at the outlet of the watershed is $t = (n-1)\theta$.

5. Discuss the characteristics of nonpoint source pollution. Why water quality modeling of nonpoint source pollution is more difficult than point source pollution?

REFERENCES

Brutsaert, W. (2005). *Hydrology: An Introduction.* Cambridge University Press, New York.

Donigian Jr, A.S., Imhoff, J.C. and Kittle Jr, J.L. (1999). HSPFParm: An interactive database of HSPF model parameters. *Version 1.0 Prepared for the U.S. EPA Office of Science and Technology.* Washington, DC.

Huang, W.L. (1937). *An Analysis of the Rainfall Runoff Correlation,* Ph.D. University of Illinois, Urbana, IL.

Linsley, R.K., Kohler, M.A. and Paulhus, J.L.H. (1982). *Hydrology for Engineers.* McGraw-Hill Book Company, New York.

Liu, C.C.K. (2007). Water quality analysis of storm runoff from Waiawa Ridge development project. *Project Completion Report Submitted to Engineering Concept, Inc.* Honolulu, HI.

Liu, C.C.K., Moravcik, P., Fernandes, K., Card, B. and Lee, T. (2012). Survey and modeling analysis of HDOT MS4 highway storm runoff on Oahu, Hawaii. *Water Resources Research Center.* University of Hawaii at Manoa, Project Report PR-2012-01.

Nash, J.E. (1959). Systematic determination of unit hydrograph parameters. *Journal of Geophysical Research,* 64(1), pp. 111–115.

National Center for Geographic Information and Analysis (NCGIA). (1990). *NCGIA Core Curriculum in GIS.* University of California, Santa Barbara, CA. Available at: https://escholarship.org/uc/spatial_ucsb_ncgia_cc [Asscessed 1 Jun. 2021].

Sherman, L.K. (1932). Stream flow from rainfall by the unit-graph method. *Engineering News Record,* 108, pp. 501–505.

US Environmental Protection Agency (USEPA). (2001). *PLOAD version 3.0: An ArcView GIS Tool to Calculate Nonpoint Sources of Pollution in Watershed and Stormwater Projects, User's Manual.* Available at: www.epa.gov/waterscience/BASINS/b3docs/PLOAD_v3.pdf [Accessed 1 Feb. 2021].

Chapter 6

River water quality modeling

6.1 EFFECTS OF RIVER HYDRAULIC PROPERTIES ON ITS SELF-PURIFICATION ABILITY

A river that receives sewage discharge is also known as the receiving river. The goal of river water pollution control is to restore and maintain the natural ecological environment of the river. To achieve this, it is necessary to establish the water quality standards according to the best use potential of each river reaches. For example, the water quality standard for drinking is much higher than that for the industrial cooling water. For the river reaches used as irrigation water, sightseeing, and leisure, there are also corresponding water quality standards. Whether a river can maintain its water quality standards after receiving pollutants is determined by the pollution discharge and the river's waste assimilative capacity, which also known as the self-purification ability. If the amount of pollution discharged into a receiving river is greater than its waste assimilative capacity, the river is polluted.

The receiving river's waste assimilative capacity is caused by the combined effects of various hydrodynamic and reaction mechanisms. The water quality model is an analytical tool that can simulate this combined effect (Figure 6.1). By inputting different pollution discharge or waste loads in the river water quality model, water quality of the receiving water can be calculated and predicted.

In natural conditions, the DO in the river is close to the saturation value, which is about 9.0 mg/L at 20°C (note: 1 mg/L is equal to 1 ppm). However, this limited DO provides indispensable condition for the survival of fish and other aquatic animals and plants. In general, when DO in river water is less than 5.0 mg/L, the growth of trout will be affected. When the DO in river water is less than 2.0 mg/L, most fishes cannot survive. If DO completely disappears and the river water becomes anaerobic, most of the aquatic animals and plants die. At this time, sulfur replaces oxygen and causes the organic matter to be anaerobically decomposed to produce stinky hydrogen sulfide.

The aerobic decomposition of the entrained organic waste will reduce the limited DO in the river water and cause serious damage to the ecological

DOI: 10.1201/9781003008491-6

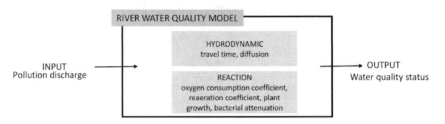

Figure 6.1 Analysis of river water quality based on comprehensive transport mechanisms.

environment. Therefore, organic waste pollution is the focus of river water pollution control in past few decades. This chapter will mainly discuss the use of water quality models to simulate the organic waste assimilative capacity of receiving rivers and to calculate the change of DO. This chapter will also discuss the simulations of the river's assimilative capacity for other traditional pollutants, including nutrients and pathogens.

In recent years, water pollution control extends to chemical fertilizers, heavy metals, pesticide residues, and other emerging pollutants. In general, the transport of organic waste in the receiving river can cause more complicated change in the DO than that of the transport of other pollutants. Therefore, the water quality model developed for organic waste simulation can be applied to the analysis of other pollutants with minor modifications.

The one-dimensional model is the most widely used analytical tool for water quality planning and management in rivers and estuaries (Fischer et al., 1979). In this chapter, the derivation of the analytical solution of the one-dimensional water quality model will be introduced, followed by the discussion of its applications. The commonly used numerical models for river water quality will also be introduced. Finally, the linear system model of river water quality, which is based on integral equations rather than differential equation, will be introduced as an alternative modeling tool.

6.1.1 Spatial distribution of pollutants in a river

The treated urban and industrial wastewater as well as storm water runoff collected by sewer systems are usually discharged into a receiving river at fixed positions, and thus, they are called point source pollution discharges (Figure 6.2). The point source pollution discharge forms a pollutant cloud or plume in the receiving river. When the polluted water flows downstream, the pollutants will mix with the surrounding water in the river. This mixing process can be divided into three stages.

The first stage is vertical mixing from the bottom to the surface, also known as initial dilution, occurring between points A and B as shown in

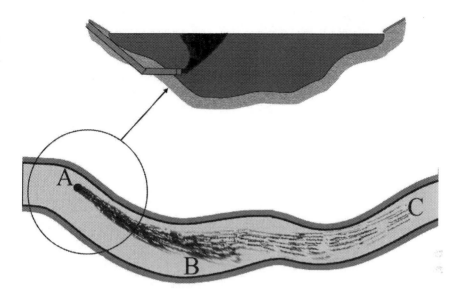

Figure 6.2 Three stages mixing process of pollutants discharged into rivers.

Figure 6.2. In this stage, a uniform distribution of contaminants in vertical direction is formed, and the initial momentum and buoyancy of the discharged water disappeared gradually. The initial dilution of urban and industrial wastewater is usually completed in a short time; therefore, the distance between A and B is small and can be ignored. In other words, it can be assumed that the concentration of pollutants in the vertical direction reaches immediately a uniform distribution state after discharge.

In coastal cities, however, the sewage is often discharged into deep seawater via marine outfalls. An example is the Sand Island Wastewater Treatment Plant in Honolulu, Hawaii. Nearly 80% of the municipal wastewater in Honolulu is treated in this treatment plant and discharged into the seawater 80 m deep in the open sea. Because the density of urban sewage is lower than that of seawater, it will rise to the sea surface after discharge. During the rising process, the initial momentum gradually disappears, and the pollutant concentration decreases. For the marine discharge of sewage, as the initial dilution takes much longer time than that in a shallow river, the analysis of the first stage of mixing becomes an important issue. A related discussion can be found in the study by Liu and Guo (1994).

The second mixing stage is the lateral mixing from one bank to the other, occurring between points B and C in Figure 6.2. The length of the distance between B and C depends on the advection effect of the river flow speed and the lateral diffusion by river turbulence. This mixing process will be discussed in detail in Section 6.1.2.

The third mixing stage is the longitudinal mixing as the pollutant plume flows downstream. The one-dimensional water quality model is the most commonly used river water quality analysis tool. Theoretically, the one-dimensional model can only simulate the longitudinal mixing of pollutants, i.e., the river reach after the completion of lateral mixing downstream of point C in Figure 6.2.

The concentration of conservative contaminants in the receiving river water is gradually diluted in the three mixing stages, which is related to the turbulent diffusion coefficient ε_t of the river (refer to Section 2.3.1) that can be estimated by the following formula:

$$\varepsilon_t = c_t h u_* \tag{6.1}$$

where c_t is a constant, h is the average water depth, and u_* is bed shear velocity. The bed shear velocity can be defined as follows: $u_* = \sqrt{\tau_0 / \rho}$, where τ_0 is the bed shear stress of the river and ρ is the river water density. The value of shear velocity in river hydraulics can be estimated using several common hydraulic parameters, for example, $u_* = \sqrt{g h s_r}$, where s_r is the longitudinal slope of the river.

6.1.2   Lateral turbulent diffusion

Based on the measured data of lateral diffusion for natural rivers with curved and irregular boundaries, the value for the constant c_t in Eqn. (6.1) is between 0.4 and 0.8 (Fischer et al., 1979, pp. 109–112). Taking the average value 0.6, the turbulent diffusion coefficient of the measured surface can be written as follows:

$$\varepsilon_t = 0.6 \, h u_* \tag{6.2}$$

The turbulent diffusion coefficient calculated by Eqn. (6.2) is only a rough estimate. A more accurate diffusion coefficient can be obtained by monitoring the tracer in the river survey. Figure 6.3 shows the lateral mixing process after the tracer is discharged from the center of the river (Fischer et al., 1979). The distance of the tracer water column flow x' is expressed as a dimensionless distance parameter $x' = (x/u)(\varepsilon_t/w^2)$. This parameter is related to the river mean flow velocity u, the river width w, and the lateral mixing coefficient ε_t.

As the tracer water column flows downstream, the concentration of the tracer gradually decreases at the center while gradually increases on both sides. Figure 6.3 shows that when $x' = 0.1$, C/C_0 approaches to 1, indicating that the concentration of pollutants at the center point

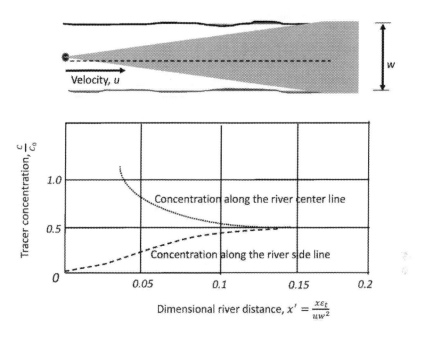

Figure 6.3 Lateral mixing of waste substances in a receiving river.

and both sides of the river is approximately the same. So practically, the dimensionless distance parameter $x' = 0.1$ can be used to estimate the flow distance of the contaminated water column required to achieve lateral mixing. The actual flow distance x_m required is thus expressed as follows:

$$x_m = 0.1 \frac{uw^2}{\varepsilon_t} \tag{6.3}$$

If the sewage is discharged from the riverbank, the range of lateral diffusion is doubled, and the actual flow distance required is expressed as follows:

$$x_m = 0.4 \frac{uw^2}{\varepsilon_t} \tag{6.3a}$$

Example 6.1 A factory discharges sewage from a river center into a river 100 ft wide and 5 ft deep. The longitudinal slope of the river is 0.0002. The mean flow velocity of the river is 2 ft/s.

Problem: Calculate the flow distance required for the contaminant water column to achieve complete lateral mixing in the river.

Solution: Calculate the bed shear velocity:

$$u_* = \sqrt{ghs_r} = \sqrt{32.2 \times 5 \times 0.002} = 0.18 \text{ ft/s}$$

Calculate the lateral mixing coefficient:

$$\varepsilon_t = 0.6hu_* = 0.6 \times 5 \times 0.18 = 0.54 \text{ ft}^2/\text{s}$$

Calculate the flow distance required to achieve lateral mixing:

$$x_m = 0.1 \frac{u(2W)^2}{\varepsilon_t} = 0.1 \frac{(2)(100)^2}{0.54} = 3,700 \text{ ft}$$

Comments

(1) In this case, the spatial distribution of pollutants in the river section is nonuniform from the discharge point to the 3,700-ft reach of downstream, where the one-dimensional model does not apply.
(2) The aforementioned analysis assumes that there are no river bends and changes in sectional shape in the receiving river. These factors usually accelerate the lateral mixing process and increase the applicable range of the one-dimensional river water quality model.

6.1.3   Longitudinal dispersion

When the flow velocity exhibits a spatial variation with internal shear stress, this type of water flow is called the shear flow. In Section 2.3.2, we have discussed that in a vertically shear flow such as that in a river, additional dispersion may be induced that enhances the mixing process in water body. In this section, we shall have a brief review of this subject as it relates to river water quality modeling.

The shear flow velocity u varying with water depth y can be expressed as follows:

$$u(y) = \bar{u} + u'(y) \tag{6.4}$$

Taylor (1954) analyzed the experimental results and found that the diffusive transport of solute in a shear flow (F_{shear}) is related to the gradient of solute concentration, similar to molecular diffusion, and can be expressed by the following formula:

$$F_{shear} = -E_x \frac{\partial C}{\partial x} \tag{6.5}$$

According to the discussion in Section 2.3.1, the one-dimensional advection–turbulent diffusion Eqn. (2.54) has the form:

$$\frac{\partial C}{\partial t} = -u\frac{\partial C}{\partial x} + \varepsilon_x \frac{\partial^2 C}{\partial x^2} \tag{6.6}$$

If Taylor's shear flow concept is considered as well for additional dispersion, Eqn. (6.6) becomes:

$$\frac{\partial C}{\partial t} = -u\frac{\partial C}{\partial x} + \varepsilon_x \frac{\partial^2 C}{\partial x^2} + E_x \frac{\partial^2 C}{\partial x^2} \tag{6.7}$$

Since the longitudinal dispersion coefficient E_x is much larger than the turbulent diffusion coefficient (Fischer et al., 1979), the equation can be reduced to:

$$\frac{\partial C}{\partial t} = -u\frac{\partial C}{\partial x} + E_x \frac{\partial^2 C}{\partial x^2} \tag{6.8}$$

Equation (6.8) is an advection–dispersion equation that simulates the transport of conservative substances in a one-dimensional river. This equation has a very similar form with the advection–turbulent diffusion equation (2.70) and the advection–molecular diffusion equation (2.54) derived in Chapter 2. Therefore, the analytical solutions to the molecular diffusion model and the turbulent diffusion model derived in the second chapter of this book is still applicable.

Assuming that the water flow in a river is in a single direction and the difference in velocity can be expressed by a logarithmic function, the approximation of the longitudinal dispersion coefficient can be estimated by using the following formula (Fischer et al., 1979):

$$E_x = 0.011\frac{u^2 W^2}{hu_*} \tag{6.9}$$

The actual distribution of river velocity is quite complex and is difficult to be represented by any specific mathematical function. Therefore, the longitudinal dispersion coefficient estimated by Eqn. (6.9) is only applicable to preliminary analyses. In Chapter 7, we will introduce the tracer method that uses the measured data from a river to determine its longitudinal dispersion coefficient.

In a one-dimensional river water quality modeling, the velocity at which a polluted water plume flows downstream is usually expressed by the time of travel. The time of travel is an important hydrodynamic parameter for understanding the nature of the river's advection property.

If the pollutant dissolved in the river water is conservative and neutrally buoyant, the time of travel θ_t in the steady flow situation is expressed as follows:

$$\theta_t = \frac{x}{u} = \frac{V}{Q} \qquad (6.10)$$

where x is the distance between the two sections of the downstream after the completion of the lateral mixing.

The time of travel of nonconservative substances in river is affected by the longitudinal dispersion and the reaction. According to the one-dimensional advection–diffusion reaction equation, the time of travel of the pollutants that are decayed by a first-order reaction in the river is calculated as follows:

$$\theta_{tc} = \theta_t \frac{1}{\sqrt{1 + \dfrac{4E_x k}{u^2}}} \qquad (6.11)$$

where θ_{tc} is the corrected travel time.

When suspended matters or other pollutants with different densities from that of river water exist in the river, the concentration will be affected by settling and other effects. In this case, the time of travel calculated by using Eqn. (6.10) or (6.11) shall be further corrected. In Section 7.2.2, the use of tracer method to measure the in situ time of travel will be introduced.

In an estuary area and a river section affected by tides, the flow rate changes periodically, so there is no direct relationship between the flow rate Q and the travel time θ_t. The flushing time is used in the estuary area instead of travel time. The definition of flushing time is expressed as follows:

$$\theta_f = \frac{V_t}{Q_f} \qquad (6.12)$$

where θ_f is the flushing time, V_t is the volume of river water relative to the average water depth between the two sections in the estuary, and Q_f is the net river flow to the downstream. The calculation of flushing time is a major task in the analysis of water quality models in the estuary area and will be discussed in detail in Section 9.1.3.

6.2 EFFECTS OF RIVER REACTION KINETICS ON ITS SELF-PURIFICATION ABILITY

The transport of nonconservative pollutants in one-dimensional rivers can be modeled by the following equation:

$$\frac{\partial C}{\partial t} = -u \frac{\partial C}{\partial x} + E \frac{\partial^2 C}{\partial x^2} + r \qquad (6.13)$$

In order to represent the natural reactions that affect river water quality in terms of reaction rate r, it is necessary to conduct experiments on individual reactions using ideal reactors and analyze the results (see Chapter 4). Various reactions related to river water quality, especially the DO, and the estimation of reaction coefficients are discussed in the following sections.

If the reaction of pollutants in river water is only of a simple first-order decay, Eqn. (6.13) can be simplified as follows:

$$\frac{\partial C}{\partial t} = -u\frac{\partial C}{\partial x} + E\frac{\partial^2 C}{\partial x^2} - kC \qquad (6.14)$$

6.2.1   Depression of river dissolved oxygen content by waste deoxygenation

The dissolved chemical matters in water can escape into the atmosphere through air–water interface; at the same time, the chemicals in the atmosphere can also enter and dissolve into water through air–water interface. When the two-way exchange rate is the same for a chemical, the concentration of the chemical in water and in atmosphere will not change and reaches the dynamic equilibrium. According to Henry's law, when the dynamic equilibrium is reached, the ratio of chemical concentration on both sides of the interface can be expressed by Henry's law:

$$\chi = K_h p_c \qquad (6.15)$$

where χ is the mole fraction of the chemical (i.e., DO) in water and p_c is the partial pressure of the chemical in atmosphere. Note that some books use different dimensional units, and the value of Henry's constant can be different from the aforementioned table. Under natural conditions, the DO saturation in a river can be calculated using Henry's law as shown in the following example.

Example 6.2 Oxygen in water and air exchange through the interface. When the rate of oxygen escaping from water to air is equal to the rate of oxygen merging into water, the exchange process reaches dynamic equilibrium. At this time, the amount of dissolved oxygen in water is called saturated dissolved oxygen, and its value can be calculated by the formula of Henry's law.

Problem: Calculate the saturated DO at 20°C using Henry's law.

Solution: Table 6.1 indicates that Henry's law constant of 20°C is 0.0000244 atm⁻¹. Because the amount of oxygen in the atmosphere is about 21%, the partial pressure of oxygen at the atmosphere is 0.21 atm. The mole fraction of oxygen in water according to point source pollution discharges is presented in Table 6.1.

$$\chi = K_h p_c = 0.0000244 \times 0.21 = 0.51 \times 10^{-5} \text{mole fraction}$$

Table 6.1 Henry's law constants of oxygen

Temperature (°C)	Henry's Law Constants Kh (atm⁻¹)
0	0.0000391
5	0.0000330
10	0.0000303
15	0.0000271
20	0.0000244
25	0.0000222
40	0.0000188
60	0.0000159

Considering a liter of water with saturated DO, the mole number of water is (1000 g)/(18 g/mole) = 55.6 mole, and the mole number of the DO is defined as C_s. The mole fraction of oxygen in water at 20°C is then:

$$\frac{C_s}{(C_s + 55.6)} = 0.51 \times 10^{-5}$$

By solving the aforementioned equation, it can be obtained that the saturated DO in 1 L of water at 20°C is $C_s = 2.84 \times 10^{-4}$ mole.

Convert this to concentration of g/m^3:

$$C_s = 2.84 \times 10^{-4} \text{ mole/L } (32 \frac{g}{mole})(10^3 \frac{mg}{g}) = 9.09 \text{ mg/L} = 9.09 \text{ g/m}^3$$

Comments

The DO content in water is mainly determined by the oxygen transfer crossing the air–water interface. When the rate of oxygen escaping from water to air is equal to the rate of oxygen dissolving from air to water, the transfer process reaches dynamic equilibrium. At this time, the DO content in water is called saturated DO, and its value can be described by Henry's law and varies with water temperature and salinity. Table 6.2 shows that the saturation value of DO in water varies with water temperature and salinity. The amount of saturated DO is about 14.6 mg/L in fresh water at 0°C and decreases to 9.09 mg/L at 20°C. In Table 6.2, the salinity is expressed by the chloride concentration in water. Seawater has a chloride concentration of about 20,000 mg/L, and the water with the chloride concentration less than 200 mg/L is called fresh water. Brackish water has chloride levels between 200 and 20,000 mg/L.

Because of the close relationship between organic waste and DO, BOD is used to express organic waste in environmental engineering. BOD is the DO consumed by microbes to decompose organic waste in water. The decomposition of organic carbon and organic nitrogen in sewage consumes the DO in water, so the BOD includes two parts: CBOD and NBOD.

Table 6.2 Saturated dissolved oxygen with temperature and salinity

Temperature (°C)	Chloride Concentration (1000 mg/L)					
	0	5	10	15	20	25
0	14.621	13.726	12.885	12.096	11.356	10.660
1	14.216	16.354	12.544	11.782	11.068	10.396
2	13.830	12.999	12.217	11.482	10.792	10.143
3	13.461	12.659	11.904	11.195	10.528	9.900
4	13.108	12.334	11.605	10.920	10.275	9.668
5	12.771	12.023	11.319	10.656	10.032	9.445
6	12.448	11.726	11.045	10.404	9.800	9.231
7	12.139	11.441	10.782	10.161	9.577	9.025
8	11.843	11.167	10.530	9.929	9.362	8.828
9	11.560	10.906	10.288	9.706	9.157	8.639
10	11.288	10.654	10.056	9.492	8.959	8.456
11	11.027	10.413	9.834	9.286	8.769	8.281
12	10.777	10.182	9.620	9.089	8.587	8.113
13	10.537	9.960	9.414	8.898	8.411	7.950
14	10.306	9.746	9.216	8.715	8.242	7.794
15	10.084	9.540	9.026	8.539	8.079	7.643
16	9.870	9.342	8.843	8.370	7.922	7.498
17	9.665	9.152	8.666	8.206	7.770	7.358
18	9.467	8.968	8.496	8.048	7.624	7.223
19	9.276	8.791	8.332	7.896	7.483	7.092
20	9.092	8.621	8.173	7.749	7.347	6.966
21	8.915	8.456	8.020	7.607	7.215	9.843
22	8.744	8.297	7.872	7.470	7.088	6.725
23	8.578	8.143	7.729	7.337	6.964	6.611
24	8.418	7.994	7.591	7.208	6.845	6.499
25	8.263	7.850	7.457	7.083	6.729	6.392
26	8.114	7.710	7.327	6.962	6.616	6.287
27	7.968	7.575	7.201	6.845	6.507	6.186
28	7.828	7.444	7.079	6.731	6.401	6.087
29	7.691	7.316	6.960	6.621	6.298	5.991
30	7.559	7.193	6.845	6.513	6.198	5.898

The aerobic decomposition of nitrogen compounds in waste water is also called nitrification. Nitrification usually starts after the decomposition of organic carbon compounds; thus, it is also called the second-stage oxygen-consuming reaction. In the early days, urban sewage was discharged into rivers without any treatment. The DO in rivers was quickly consumed by a large amount of carbonaceous oxygen-consuming wastes, presenting a serious pollution state. Therefore, the early analysis of river water quality model such as the Streeter–Phelps model only considers the first-stage carbonaceous oxygen-consuming reaction. BOD in the commonly used Streeter–Phelps model equation is CBOD.

6.2.1.1 First-stage carbonaceous wastes (CBOD) deoxygenation reaction

In the carbonaceous waste deoxygenation experiment, the water sample is diluted with pure water and put into several BOD bottles with a volume of 300 mL. After the solution in the bottles is completely mixed, the initial DO in the bottles is recorded. The bottles are then incubated in a refrigerator at 20°C. BOD bottles can be regarded as batch reactors as discussed in Chapter 4. After t days, start to take out these bottles one after another and record their DO, then calculate the BOD contained in water sample during t-day incubation with the following formula:

$$y(t) = \frac{DO_i - DO_t}{\gamma_d} \tag{6.16}$$

where $y(t)$ is t-day BOD consumed (mg/L), DO_i is the initial DO in the bottle (mg/L), DO_t is the DO in the bottle after t days (mg/L), and γ_d is the dilution ratio (%).

The BOD consumed $y(t)$ is a function of time. In the design of urban sewage treatment plants, BOD_5 $y(5)$ is usually used to represent the amount of organic carbon compounds, also known as BOD_5 (Figure 6.4). The BOD_5 in untreated urban sewage is about 200 mg/L, which is much larger than the saturated DO in the water. This is the reason why dilution of the water sample is usually needed in the BOD determination experiments.

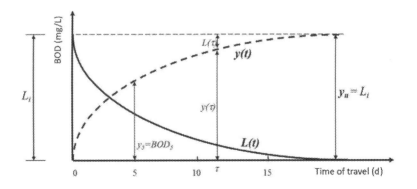

Figure 6.4 Relationship between BOD consumed y(t) and BOD remaining L(t).

There are two different definitions of BOD: BOD consumed and BOD remaining. Figure 6.4 shows the relationship between them. The $y(t)$ calculated by Eqn. (6.16) is the BOD consumed after t-day. When a water sample is incubated in a BOD bottle for a long period of time, the organic waste in the sample has been completely consumed by microorganisms, and the BOD consumed is called ultimate BOD (y_u or BOD_u). The ultimate BOD is a constant, which is also equal to the BOD remaining at the beginning of the BOD deoxygenation experiment, L_i. As shown in Figure 6.4, $y_u = L_i$. It can also be seen from Figure 6.4 that during the experiment, the BOD consumed and the BOD remaining change with time, but at any given time, the sum of the two is equal to the ultimate BOD, which is a constant. It is important to understand these two different definitions of BOD, as the BOD, L in the water quality model equations refers to the ultimate BOD_u, while the design and operation of sewage treatment plants generally use the amount of DO consumed for 5-d or BOD_5.

The oxygen-consumption reaction of organic waste can be expressed by the following first-order reaction equation:

$$r_C = \frac{dL}{dt} = -k_c L \tag{6.17}$$

where k_c is the first-order deoxygenation coefficient of the carbonaceous wastes, d^{-1}. When the initial conditions are: $L = L_0$ at $t = 0$, the solution of Eqn. (6.17) is as follows:

$$L(t) = L_0 \exp(-k_c t) \tag{6.18}$$

Example 6.3 The discharge water quality of a secondary sewage treatment plant is analyzed in a laboratory, and find that the BOD_7 and BOD_u are 10 mg/L and 50 mg/L, respectively.

Problem: Calculate (1) deoxygenation coefficient k_c and (2) BOD_5.

Solution: At any given time, the sum of the BOD consumed, and the BOD remaining is equal to the ultimate BOD:

$$L(7) + y(7) = L_0$$

Therefore, the 7-d BOD is expressed as follows:

$$L(7) = L_0 - y(7) = 50 - 10 = 40 \text{ mg/L}$$

(1) Calculate the deoxygenation coefficient, k_c:

Taking the logarithms of Eqn. (6.18), we have the following relationship:

$$k_c = \frac{-(\ln L_t - \ln L_0)}{t} \tag{6.19}$$

Substituting the known 7-d oxygen demand $L = (7) = 40$ mg/L and the ultimate BOD $L_0 = 50$ mg/L into Eqn. (6.19) gives:

$$k_c = \frac{-\ln\left(\dfrac{L_7}{L_0}\right)}{7} = \frac{-\ln\left(\dfrac{40}{50}\right)}{7} = 0.23 \text{ d}^{-1}$$

(2) Calculate BOD$_5$ or $y(5)$:

Use Eqn. (6.18) to calculate BOD$_5$ remaining, $L(5)$:

$$L(5) = L_0 \exp\left[-k_c(5)\right] = 50 \exp[-0.23(5)] = 15.8 \text{ mg}/L$$

Calculate the BOD consumed in 5 d:

$$\text{BOD}_5 = y(5) = L_0 - L(5) = 50 - 15.8 = 34.2 \text{mg}/L$$

The deoxygenation coefficient of river water changes with temperature. The reaction in the BOD bottle is performed at a temperature of 20°C. When the obtained deoxygenation coefficient is used at other temperatures, the following equation should be used to adjust the value:

$$\left(k_c\right)^t = \left(k_c\right)^{20}\left(f_{tc}\right)^{t-20} \tag{6.20}$$

where $(k_c)^t$ is the oxygen consumption coefficient at the temperature t °C, $(k_c)^{20}$ is the oxygen consumption coefficient at the temperature of 20°C, f_{tc} is the temperature adjustment factor, and the generally adopted value is $f_{tc} = 1.05$ (Liu, 1986).

In a real river, the deoxygenation coefficient is also affected by river morphology and hydrodynamics and can be adjusted using the following formula:

$$k_d = k_c + f_n\left(\frac{u}{h}\right) \tag{6.21}$$

where k_d is the river deoxygenation coefficient, d^{-1}, and f_n is the oxygen consumption factor of the river bed (Table 6.3).

Table 6.3 Factors affecting oxygen consumption rate under different riverbed slopes

Riverbed Slopes (mile)	Oxygen Consumption Factor (fn)
2.5	0.10
5.0	0.15
10.0	0.25
25.0	0.40
50.0	0.60

The oxygen consumption coefficient obtained by using Eqn. (6.21) is an estimation. In Chapter 7, we will discuss how to carry out in situ water quality measurement to obtain the river's deoxygenation coefficient.

6.2.1.2 Nitrification and second-stage deoxygenation reaction

Organic nitrogen in wastewater is converted into ammonia nitrogen, nitrite nitrogen, and nitrate nitrogen by a series of oxidation by oxygen-consuming nitrifying bacteria. Figure 6.4 shows deoxygenation reactions of carbonaceous oxygen-consuming substances in wastewater, that is, there is no nitrogenous oxygen-consuming substances in the water sample, or a nitrification inhibitor is added in the oxygen consumption experiment.

The experiment of oxygen consumption by both carbonaceous and nitrogenous substances is conducted in laboratory and shown in Figure 6.5. The nitrification in wastewater usually starts when the decomposition of carbonaceous wastes is almost completed. Thus, it is called the second-stage deoxygenation reaction. According to the stoichiometric relationship, 4.57 mg/L of DO is needed to decompose 1 mg/L of organic and ammonia nitrogen whose total amount is called TKN (total Kjeldahl nitrogen). The theoretical value of NBOD in a water sample can be calculated using the following formula:

$$(L_n)_0 = 4.57 (\text{TKN}) \tag{6.22}$$

where $(L_n)_0$ is the second-stage BOD or NBOD (mg/L).

The early river water quality modeling only considered the oxygen consumption reaction of carbonaceous wastes or first-stage deoxygenation. In

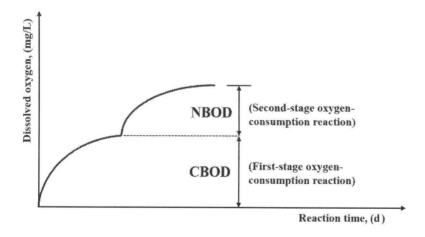

Figure 6.5 Second-stage oxygen-consumption reaction.

recent years, secondary sewage treatment plants which can remove more than 85% of carbonaceous wastes from sewage have been widely built and used. Therefore, the further water quality improvement must consider the second-stage deoxygenation. The reaction rate equation is as follows:

$$r_n = \frac{dL_n}{dt} = -k_n L_n \tag{6.23}$$

where k_n is the second-stage deoxygenation coefficient, or the nitrification coefficient (d^{-1}):

The nitrification coefficient is related to temperature:

$$(k_n)^t = (k_n)^{20} (f_{tn})^{t-20} \tag{6.24}$$

where $(k_n)^t$ is the nitrification coefficient at temperature t (°C), $(k_n)^{20}$ is the nitrification coefficient at temperature 20°C, and f_{tn} is a temperature adjustment factor, with the generally used value of 1.08 (Liu, 1986).

When the initial conditions are as follows: $L = (L_n)_0$ at $t = 0$, the analytical solution of Eqn. (6.23) is given as follows:

$$L_n(t) = (L_n)_0 \exp(-k_n t) \tag{6.25}$$

Example 6.4 Water samples taken from the sewage treatment plant are sent to the laboratory for long-term oxygen consumption analysis. The analysis results are as follows (Figure 6.6):

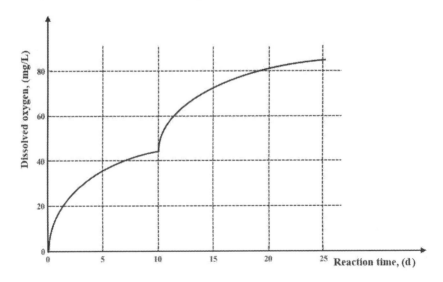

Figure 6.6 Oxygen-consumption curve of water sample from a sewage treatment plant.

Problem: Calculate (1) BOD$_5$, (2) CBOD, (3) NBOD, and (4) CBOD oxygen consumption coefficient k_c.

Solution: The answers to the first three questions can be obtained directly from Figure 6.6:

(1) BOD$_5$ = y_5 = 40 mg/L
(2) CBOD = 46 mg/L
(3) NBOD = 85 − 46 = 39 mg/L
(4) The fourth problem can be solved according to Eqn. (6.18). The following two points are selected in the first-stage oxygen-consuming curve:

$$t_a = 2 \text{ d}, \; L_a = L_u - y_a = (46 - 21) = 25 \text{ mg/L}$$
$$t_b = 5 \text{ d}, \; L_b = L_u - y_b = (46 - 40) = 6 \text{ mg/L}$$
$$k_c = \frac{\ln(25) - \ln(6)}{(5-2)} = 0.48 \text{ d}^{-1}$$

Comments

(1) The untreated municipal sewage contains about 200 mg/L of BOD$_5$, and the wastewater after secondary treatment should contain less than 30 mg/L of BOD$_5$. Therefore, the sewage treatment plant in the example does not meet the secondary treatment requirements.
(2) From the experimental data, it can be seen that the oxygen consumption reaction of the organic carbon in the first stage ends in about the 10th day, and the nitrification in the second stage ends in about the 25th day. These two time points will vary slightly depending on the nature of the wastewater. Generally, the experimental period in a laboratory analysis of the oxygen consumption of sewage samples is 28 d, and the 28-d oxygen consumption is taken as the total oxygen demand (TOD), TOD = CBOD + NBOD.

The carbonaceous organic wastes (CBOD, L_C) may exist in sewage in either dissolved form or attached to the surface of suspended particles. When sewage is discharged into rivers, the CBOD in sewage will decrease due to the oxygen-consuming decomposition and the settlement of suspended particles. If the settlement of suspended particles can also be considered as a first-order reaction, the removal rate of oxygen-consuming carbonaceous organic wastes in the receiving river can be expressed by the following formula:

$$k_r = k_s + k_c \tag{6.26}$$

where k_r is the removal coefficient of carbonaceous organic wastes in a river reach and k_s is the settling coefficient of carbonaceous organic wastes. Field measurements of these coefficients are discussed in Section 7.3.1.

Table 6.4 The value of benthic uptake coefficient (20°C; Thomann, 1972)

Type of Sludge	Benthic Uptake Coefficient (g/m²/d)	
	Numeric Ranges	Average Value
Sphaerotilus (10 g dry wt/m²)	–	7.0
Municipal sewage sludge-outfall vicinity	2.0–10.0	4.0
Municipal sewage sludge – 'aged' downstream of outfall	1.0–2.0	4.0
Cellulosic fiber sludge	4.0–10.0	7.0
Estuarine mud	1.0–2.0	1.5
Sandy bottom	0.2–1.0	0.5
Mineral soils	0.05–0.1	0.07

The benthal deposits are rich in oxygen-consuming materials due to the settling of organic wastes. The benthic uptake refers to the absorption and utilization of DO when it enters the sediment and is consumed by the organic matter. In addition, the organic waste settled in the bottom mud is also washed back to the water and then the oxygen-consuming reaction takes place in the water. In the analysis of river water quality models, these two types of benthic uptake are usually considered as another form of non-point source pollution, and the following zero-order reaction formula is used to simulate the reaction:

$$r_B = -S_B \qquad\qquad (6.27)$$

where r_B is the benthic uptake rate (g/m²/d) and S_B is the benthic uptake coefficient (g/m²/d). Dividing S_B by the water depth converts the coefficient into the benthic uptake coefficient (g/m³/d) relative to the volume of river water.

The value of the benthic uptake coefficient suggested in the literature varies widely from 0.1 to 10.0 g/m²/d. Table 6.4 was proposed by Thomann (1972) based on the literature results, and its values are often cited. Like other model coefficients, the ideal method is to carry out field measurements in the investigated river reaches and use the measured data to estimate the value of benthic uptake coefficient (see Chapter 7).

6.2.2   Replenishment of dissolved oxygen content by reaeration

When the amount of the DO in the river is lower than the saturation level, the oxygen in the air will enter the river through the interface transfer. The interface transfer of oxygen between air and river water is the reaeration

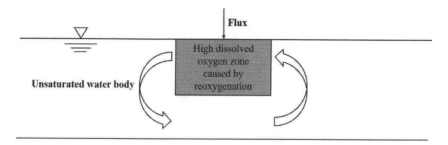

Figure 6.7 Illustration of reaeration in a river.

phenomenon (Figure 6.7). The interface transfer is an important topic in environmental engineering, which occurs between liquid–solid, liquid–gas, and solid–gas. In addition to affecting the self-purification ability of water environment, the principle of interface transfer is also used to design various sewage treatment methods including activated carbon adsorption of liquid–solid interface transfer and gas stripping of liquid–air–liquid transfer.

The reaeration phenomenon can be expressed by the following first-order reaction equation:

$$r_{reaeration} = \frac{dC}{dt} = k_2 (C_s - C) \tag{6.28}$$

where $C_S - C$ is the difference of the DO from the DO saturation and is defined as DO deficit D, and k_2 is the reaeration coefficient.

When the oxygen in the air enters a river, the DO on the surface layer of the river will soon become saturated, that is, $C = C_S$ (Figure 6.7), and the reaeration will stop. Therefore, whether the reaeration can continue and how fast it can be are determined by the turbulence intensity of a river that affects the exchange rate of DO between surface and interior river water.

There are two difficulties in estimating the value of river reaeration coefficient: (1) The measured DO in river is affected by oxygen consumption, reaeration, benthic uptake, and the photosynthesis/respiration of plants in water. Therefore, the observed change of DO in river cannot be directly used to estimate the reaeration coefficient. (2) The effect of turbulence intensity on river reaeration is difficult to model in laboratory experiments. Currently, empirical formulae are often used to calculate the reaeration coefficient in the river water quality modeling. These formulae are obtained by observing the change in DO under the reaeration in the artificial channel in the laboratory, and several hydraulic parameters including water depth, velocity, and hydraulic gradient of riverbed are used as indicators of river turbulence

Table 6.5 Empirical formulas for estimating reaeration coefficient

Formula	Field Conditions	Reference
$k_2 = 3.93 \dfrac{V^{0.5}}{\overline{h}^{-1.5}}$	$0.3\text{ m} < \overline{h} < 9\text{ m},$ $0.15\text{ m/s} < V < 0.5\text{ m/s}$	O'Connor and Dobbins (1958)
$k_2 = 5.23 \dfrac{V}{\overline{h}^{-1.67}}$	$0.6\text{ m} < \overline{h} < 3\text{ m},$ $0.55\text{ m/s} < V < 1.50\text{ m/s}$	Churchill et al. (1962)
$k_2 = 5.32 \dfrac{V^{0.67}}{\overline{h}^{-1.85}}$	$0.1\text{ m} < \overline{h} < 3\text{ m},$ $0.03\text{ m/s} < V < 1.50\text{ m/s}$	Owens et al. (1964)
$k_2 = 3.1 \times 10^4 V S_0$	$0.3\text{ m} < \overline{h} < 0.9\text{ m},$ $0.03\text{ m}^3\text{/s} < Q < 0.3\text{m}^3\text{/s}$	Tsivoglou and Wallace (1972)

intensity. Table 6.5 lists the most commonly used empirical formulas and their application ranges.

6.2.3   Dissolved oxygen variation in biologically active rivers

Carbon, oxygen, hydrogen, potassium, nitrogen, and phosphorus are all essential elements for plant growth. In addition, other trace elements such as iron, calcium, and copper are also necessary for plant growth. When these elements are abundant in lakes and rivers, the water is eutrophic that enables aquatic plants, such as algae, to grow quickly. This limits the growth of other aquatic lives due to the lack of supply of these elements as well as the exhausting of DO. In natural lakes and rivers, the most common limiting elements for aquatic plants are nitrogen and phosphorus, which are absorbed by aquatic plants in the form of nitrates and phosphates. Therefore, nitrates and phosphates are often referred to as limiting nutrients.

Limited by nitrates or phosphates, there are few problems of eutrophication in unpolluted natural water. When these nutrients are discharged into lakes/rivers along with other pollutants, the natural restrictions are removed, and eutrophication occurs. When the surface of a lake/river is covered by aquatic plants due to eutrophication, reaeration from the atmosphere to the water body will be hindered, resulting in the decline of DO and deterioration of the ecological environment. A large number of aquatic plants can also reduce the function of lakes/rivers as public water sources and navigation. A more serious situation is the emergence of harmful algal bloom in the eutrophic lakes/rivers, such as the eutrophication and algal bloom in Taihu Lake in the 2000s (Qin et al., 2010).

The growth rate of algae in river water is limited by the concentration of nutrients, which can be explained by the saturation reaction described in Section 4.2.4. The relevant mathematical formula is as follows:

$$k_G = k_{Gmax} \frac{C_{nu}}{K_s + C_{nu}} \qquad (6.29)$$

where k_G is the plant growth rate (d^{-1}) limited by nutrients in water, k_{Gmax} is the maximum reaction rate (d^{-1}), C_{nu} is the nutrient concentration (mg/m^3), and K_s is the saturation constant (mg/m^3), which is the nutrient concentration when the plant growth rate in water is 50% of the maximum value.

As the growth rate of algae in natural water is usually limited by nitrates or phosphates, Eqn. (6.29) can be written as follows:

$$k_G = k_{Gmax} \left[\min\left(\frac{C_n}{K_s + C_n}, \frac{C_p}{K_s + C_p} \right) \right] \qquad (6.30)$$

where C_n is the concentration of nitrate (mg/m^3) and C_p is the concentration of phosphate (mg/m^3).

The photosynthesis and respiration of aquatic plants, especially algae, can affect the change of DO in rivers or other water bodies. According to O'Connor and Di Toro (1970), the diurnal variation of photosynthesis of aquatic plants affected by sunlight can be roughly expressed as a sine function, that is, the rate of $P(T)$ produced by photosynthesis can be expressed as follows:

$$P(t) = P_m \sin\left[\omega(t - t_r) \right] \qquad t_r < t < t_s \qquad (6.31)$$

where P_m is the highest value of photosynthesis in a day (g/m^{-3} d^{-1}), $\omega = \pi/fT_p$ with T_p is the period of 1 d or 24 h, and the time of sunrise and sunset is t_r and t_s, respectively.

The relationship between P_m and average value P is as follows:

$$P = \frac{\int_0^{T_p} P(t)dt}{T_p} = \frac{\int_0^{T_p} P_m \sin\left[\omega(t - t_r) \right] dt}{T_p} = P_m \frac{2f}{\pi} \qquad (6.32)$$

The respiration R of aquatic plants is usually regarded as a constant in time; thus, the change of the DO caused by photosynthesis and respiration of aquatic plants can be expressed by

$$r_{P-R} = \frac{dC}{dt} = P(t) - R \qquad (6.33)$$

6.2.4 River bacterial pollution and indicating microorganisms

The basic goal of water pollution control is to maintain the natural ecological environment and public health. An important work of public health maintenance is to control and eliminate various pathogens in receiving rivers. As the possibility of any kind of pathogens in receiving river is generally not high, the indicator bacteria are usually used in water quality management to understand the potential of pathogen pollution. The most commonly used indicator bacteria are coliform bacteria, which are generally found in water, harmless to human health, and easy to be measured. In addition, coliform bacteria are relatively stable in water environment, and their decay rate in the receiving river is slower than that of other pathogens. Hence, coliform bacteria can provide a conservative prediction.

The content of coliform bacteria in water can be determined by the membrane filtering technique, coli-count sampler, and multitube fermentation technique (Vesilind and Peirce, 2013). Coliform bacteria can ferment in Lactose Broth to produce gas and make the solution turbid. The multitube fermentation technique is to dilute the water samples to different degrees and put them into three tubes with Lactose Broth to observe whether there are coliform bacteria. Finally, the most probable number (MPN) of coliform bacteria in the original water sample is calculated by the statistical method. The coliform bacteria MPN of 100 mL of untreated domestic sewage is about 1,000,000 or 1,000,000 E. coli/100 mL. The drinking water quality standard requires that there should be less than 200 coliform bacteria in 100 mL of water or 200 E. coli/100 mL.

The decay of coliform bacteria in the receiving river is generally regarded as the first-order reaction.

$$\frac{dN_B}{dt} = -k_b N_B \tag{6.34}$$

where N_B is the number of coliform bacteria (E. coli/100 mL) and k_b is the first-order decay coefficient (d^{-1}) of coliform bacteria in the river. According to the measured results in different rivers, the value of k_b is between 0.96 and 2.88 d^{-1} (Zison et al., 1978).

6.3 SIMPLIFIED RIVER WATER QUALITY MODELS

6.3.1 The Streeter–Phelps model

The DO of river is affected by oxygen consumption of organic waste, atmospheric reaeration, benthic uptakes, and plant photosynthesis/respiration. Therefore, the reaction term r in Eqn. (6.12) can be expressed as follows:

$$r = r_C + r_N + R_{reaeration} + S_B + r_{P-R} = k_c L_c + k_n L_n - k_2 D + S_B - [P(t) - R] \tag{6.35}$$

The following one-dimensional river DO model can be obtained by using the DO deficit D as the unknown variable, or $D = C_s - DO$. Substituting Eqn. (6.35) into Eqn. (6.13):

$$\frac{\partial D}{\partial t} = -u\frac{\partial D}{\partial x} + E\frac{\partial^2 D}{\partial x^2} + k_c L_c + k_n L_n - k_2 D + S_B - [P(t) - R] \qquad (6.36)$$

There are three unknown variables D, L_c, and L_n in Eqn. (6.36). The variation of D in the river is complicated and includes the time-varying term $P(t)$ produced by the photosynthesis of plants in the water. The other two unknown variables, the reaction of oxygen-consuming carbides (CBOD, L_c) and oxygen-consuming nitrides (NBOD, L_n) in river, are simply of first-order decay and vary only with distance:

CBOD:

$$\frac{\partial L_c}{\partial t} = -u\frac{\partial L_c}{\partial x} + E\frac{\partial^2 L_c}{\partial x^2} + k_c L_c \qquad (6.37)$$

NBOD:

$$\frac{\partial L_n}{\partial t} = -u\frac{\partial L_n}{\partial x} + E\frac{\partial^2 L_n}{\partial x^2} + k_n L_n \qquad (6.38)$$

To solve the temporal and special changes of DO in the receiving river, the aforementioned three differential equations need to be solved together, that is, first, solve Eqns. (6.37) and (6.38) and then substitute the calculated L_c and L_n into Eqn. (6.36) to obtain the temporal and spatial change of DO in the receiving river. In general, these equations cannot be solved analytically. Several numerical models produced by the USEPA and the US Army Corps of Engineers are discussed in Section 6.4.

In this section, we will discuss how to introduce reasonable assumptions to simplify Eqn. (6.36) and obtain analytical solutions accordingly. First, assuming that the photosynthetic/respiration of the plant that changes with time in water can be ignored or replaced by its daily average, the unknown variable (DO deficit) in Eqn. (6.36) is just a function of distance, or $\partial D/\partial t = 0$. Further assuming the longitudinal dispersion, nitrification and benthic uptakes are negligible, the partial differential Eqn. (6.36) is simplified as the Streeter–Phelps model:

$$u\frac{dD(x)}{dx} = k_1 L(x) - k_2 D(x) \qquad (6.39)$$

Since the oxygen consumption of the nitride has been neglected, Eqn. (6.39) $L(x)$ only represents carbonaceous oxygen-consuming wastes, or CBOD, hereinafter referred to as BOD.

Equation (6.39) can be directly derived if (1) the hydrodynamic mechanism of the receiving river is the ideal PFR under the steady state; (2) the reaction mechanism consists of the decomposition of oxygen-consuming carbide and the atmosphere reaeration (refer to Section 4.3.3. for a more detailed discussion).

There are two unknown variables $D(x)$ and $L(x)$ in Eqn. (6.39). Therefore, another independent equation is needed for closure. This independent equation that simulates the change of $L(x)$ in the river is the simplified Eqn. (6.37):

$$u\frac{dL(x)}{dx} = -k_1 L(x) \tag{6.40}$$

Theoretically, the Streeter–Phelps model consists of two equations, Eqns. (6.39) and (6.40), which describe the concentration changes of DO and oxygen-consuming wastes or BOD in the receiving river. Equations (6.39) and (6.40) can also be written as equations with the time of travel as an independent variable:

$$\frac{dD(\theta_t)}{d\theta} = k_1 L(\theta_t) - k_2 D(\theta_t) \tag{6.41}$$

$$\frac{dL(\theta_t)}{d\theta_t} = k_1 L(\theta_t) \tag{6.42}$$

There are three model coefficients in Eqns. (6.41) and (6.42). While the travel time (θ_t) is the hydrodynamic coefficient, the deoxygenation coefficient k_1 and the reaeration coefficient k_2 are reaction coefficients. This simple model, similar to other complex models, has the ability to simulate hydrodynamic and reaction kinetics of river water quality systems (Figure 6.8).

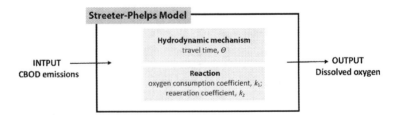

Figure 6.8 Streeter–Phelps river water quality model.

With the boundary condition: $L = L_0$ at $x = 0$, the solution of Eqn. (6.42) can be obtained directly by integration:

$$L(\theta_t) = L_0 e^{-k_1\theta_t} \tag{6.43}$$

By introducing Eqn. (6.43) into Eqn. (6.41), the DO balance equation in the Streeter–Phelps model can be written as follows:

$$\frac{dD}{d\theta_t} + k_2 D = k_1 L_0 e^{-k_1\theta_t} \tag{6.44}$$

The Streeter–Phelps model assumes that the mixing process of waste substances and river water is instantaneous; thus, the initial BOD L_0 and the initial DO content DO_0 (or the initial DO difference D_0) can be easily calculated by a simple law of conservation of mass.

The ordinary differential equation (6.44) can be solved by the integral factor method (see Sokolnikoff and Redheffer, 1966). The integral factor of Eqn. (6.44) is $e^{\int k_2 d\theta_t} = e^{k_2\theta_t}$. By multiplying the factor into the equation, we have:

$$\frac{dD(e^{k_2\theta_t})}{d\theta_t} = k_1 L_0 e^{(k_2-k_1)\theta_t}$$

After integrating this equation, we obtain:

$$D(e^{k_2\theta_t}) = \frac{k_1 L_0}{k_2 - k_1} L_0 e^{(k_2-k_1)\theta_t} + c$$

The integral constant c can be determined by applying the boundary condition: $\theta_t = 0$, $D = D_0$:

$$c = D_0 - \frac{k_1 L_0}{k_2 - k_1}$$

Substituting this integral constant, the solution of DO in the river with the travel time θ_t is expressed as follows:

$$D(\theta_t) = \frac{k_1 L_0}{k_2 - k_1}\left(e^{-k_1\theta_t} - e^{-k_2\theta_t}\right) + D_0 e^{-k_2\theta_t} \tag{6.45}$$

Because the time of travel is $\theta_t = \dfrac{x}{u}$, the solution of BOD and DO in rivers with distance x:

$$L(x) = L_0 e^{-k_1\frac{x}{u}} \tag{6.46}$$

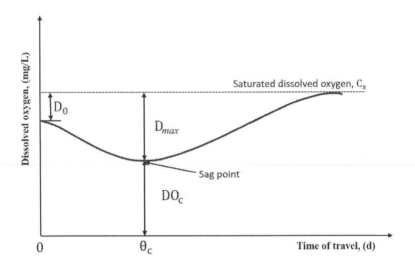

Figure 6.9 Sag curve of DO in a receiving river.

$$D(x) = \frac{k_1 L_0}{k_2 - k_1} (e^{-k_1 \frac{x}{u}} - e^{-k_2 \frac{x}{u}}) + D_0 e^{-k_2 \frac{x}{u}} \qquad (6.47)$$

Equation (6.45) describes the variation curve of DO concentration with the time of travel in the downstream reach after sewage is discharged into the receiving river. The curve is often called the sag curve because of its shape (Figure 6.9).

The DO in the unpolluted river water should be saturated. Figure 6.9 shows that at the discharge point ($\theta_t = 0$), the DO is slightly below the saturation value after mixing with the sewage water. The BOD is large near the discharge point, so the deoxygenation rate is large. At the same time, the DO deficit is small, and the reaeration rate is small. The DO gradually decreases when the deoxygenation rate is much larger than the reaeration rate. As the river continues to flow downstream, the BOD gradually decreases, while the DO deficit increases. Therefore, the deoxygenation slows down, and the reaeration accelerates. When the deoxygenation rate equals reaeration rate, the sag curve reaches the lowest point, also known as the sag point. Downstream from this point, the deoxygenation rate is less than the reaeration rate, and the DO content gradually recovers and eventually returns to the saturated state.

From the view of water quality management, the sag point is the most important consideration. When the DO content at the sag point is above the water quality standard, it means that the sewage discharge does not cause pollution problems in the river. On the contrary, when the DO content at the sag point is below the water quality standard, it means that it is necessary to reduce the pollutant loads.

The sag point occurs when the first differential of the sag curve equation is equal to zero:

$$\frac{dD}{d\theta_t} = 0 = \frac{k_1 L_0}{k_2 - k_1}\left(-k_1 e^{-k_1 \theta_t} + k_2 e^{-k_2 \theta_t}\right) - k_2 D_0 e^{-k_2 \theta_t} \tag{6.48}$$

The equation for calculating the position of the sag point is obtained by rearranging the above equation:

$$\theta_c = \frac{1}{k_2 - k_1} \ln\left[\frac{k_2}{k_1}\left(1 - D_0 \frac{k_2 - k_1}{k_1 L_0}\right)\right] \tag{6.49}$$

In addition, the minimum DO content occurs when the rate of the oxygen-consuming reaction is equal to the rate of the reaeration:

$$k_1 L_c = k_2 D_{max} \tag{6.50}$$

Substituting θ_c calculated by Eqn. (6.49) into Eqn. (6.50), the maximum DO deficit is obtained:

$$D_{max} = \frac{k_1}{k_2} L_c = \frac{k_1}{k_2} L_0 e^{-k_1 \theta_c} \tag{6.51}$$

The minimum DO DO_c can be obtained:

$$DO_c = C_s - D_{max} = C_s - \frac{k_1}{k_2} L_0 e^{-k_1 \theta_c} \tag{6.51a}$$

Example 6.5 A municipal WWTP discharges 1.5 m³/s of sewage into a nearby river. The river flow rate is 4 m³/s. The DO after mixing water with discharged sewage is 7.5 mg/L, and the temperature of both the river water and sewage is 25°C. The oxygen consumption coefficient is 0.2 d⁻¹, and the reaeration coefficient of the river section below the discharge point is 0.41 d⁻¹. The DO of the river must be maintained above 5.5 mg/L.

Problem: Calculate the maximum allowable BOD of the discharged sewage.

Solution: The content of saturated DO at a water temperature of 25°C is $C_s = 8.2$ mg/L (Table 6.2). Consider the standard of DO of the river water is 5.5 mg/L, and the maximal DO deficit is expressed as follows:

$$D_{max} = 8.2 - 5.5 = 2.7 \, \text{mg}/\text{L}$$

The oxygen consumption coefficient is measured at 20°C, and it is expressed as follows:

$$(k_1)_t = (k_1)_{20} - (1.05)^{25-20} = 0.2(1.05)^5 = 0.26 \, \text{mg}/\text{L}$$

According to Eqn. (6.50):

$$\theta_c = \frac{1}{k_2 - k_1} \ln\left[\frac{k_2}{k_1}\left(1 - D_0 \frac{k_2 - k_1}{k_1 L_0}\right)\right]$$

$$\theta_c = \frac{1}{0.41 - 0.26} \ln\left[\frac{0.41}{0.26}\left(1 - 0.7\frac{0.41 - 0.26}{0.26 L_0}\right)\right]$$

(a)

According to Eqn. (6.51):

$$D_{max} = \frac{k_1}{k_2} L_c = \frac{k_1}{k_2} L_0 e^{-k_1 \theta_c}$$

$$2.7 = \frac{0.26}{0.41} L_0 \exp(-0.26\theta_c)$$

(b)

θ_c in (a) and (b) can be solved by the trial and error (Table 6.6). First, assume a θ_c value, substitute the θ_c into (a) to calculate L_0, and then substitute this L_0 into (b) to calculate θ_c. The iteration stops when the two consecutive results are close enough.

The answer is: θ_c = 2.71. Then, use formula (b) to calculate L_0 = 8.59 mg/L.

Finally, the maximum allowable BOD of the discharged sewage is calculated by the law of conservation of mass (assuming that there is only a trace amount of BOD in the upstream of the river).

$$BOD_n = \frac{5.5}{1.5}(8.59) = 31.5 \, \text{mg} / \text{L}$$

When using the Streeter–Phelps model for water quality modeling, the BOD is expressed by the ultimate BOD, BOD_u, which represents the DO content used in the decomposition of all organic wastes that can be decomposed by microorganisms.

Table 6.6 Trial and error

θ_c (Assumed Value, d)	θ_c (Calculated Value, d)
2.0	2.65
2.25	2.67
2.50	2.69
2.71	2.71
2.75	2.72
3.00	2.74

In the design and operation of WWTPs, BOD is usually expressed as BOD consumed for 5 days, or BOD_5. Using Eqn. (6.18), the BOD remaining $L(5)$ can be calculated as follows:

$$L(5) = BOD_n \exp(-k_1 t) = 31.5 \exp[-(0.26)(5)] = 8.6 \text{ mg/L}$$

Because the BOD consumed plus the BOD remaining is equal to the ultimate BOD, we then have:

$$BOD_5 = BOD_n - L(5) = 31.5 - 8.6 = 22.9 \text{ mg/L}$$

Comments

The modeling results indicate that in order to meet receiving water quality standards, the BOD_5 in the effluent from the city WWTP should be less than 22.9 mg/L. The effluent from a secondary WWTP contains BOD_5 of 30 mg/L. In order to meet the receiving river's DO standard, this WWTP must provide better than secondary treatment.

6.3.2 The modified Streeter–Phelps models

After the 1970s, the river water quality model became an indispensable analysis tool for designing large-scale sewage treatment plants. The huge investment of a large sewage treatment plant requires better accuracy of modeling. To improve the Streeter–Phelps model that has been widely used for many years is necessary.

To improve the Streeter–Phelps model, it is necessary to reduce the number of assumptions introduced during the derivation, which include the following: (1) the receiving river is a one-dimensional system, (2) the receiving river is in a steady state, and the DO varies only with distance (or travel time), (3) the advection transmission is much larger than the diffusion transmission so that the diffusion term can be ignored, and (4) only two natural reactions, carbide oxygen-consuming and atmospheric reaeration, are considered.

The following paragraphs will discuss the addition of more hydrodynamic and reaction simulation mechanisms back to the original Streeter–Phelps model equation to extend its application range and their rationality.

6.3.2.1 Enhancement of hydrodynamic mechanism simulation – addition of dispersion term

The traditional Streeter–Phelps model assumes that the longitudinal dispersion coefficient is zero. After considering the longitudinal dispersion, the model governing Eqn. (6.39) that simulates the change of DO (expressed by the DO deficit) is expanded to:

$$u\frac{dD(x)}{dx} = -E\frac{d^2 D}{dx^2} + k_1 L(x) - k_2 D(x) \tag{6.52}$$

The model governing Eqn. (6.40) for simulating BOD changes is expanded to:

$$u\frac{dL(x)}{dx} = E\frac{d^2L(x)}{dx^2} - k_1 L(x) \tag{6.53}$$

Let's first solve the enhanced BOD model or Eqn. (6.53) and discuss its rationality.

Equation (6.53) is a linear ordinary differential equation in which the coefficients are constants, and thus, the analytical solution can be obtained. First, let's set L equal to e^{nt} and substitute L into Eqn. (6.53) and convert the equation to an algebraic equation. The final solution reads (Fischer et al., 1979):

$$L(x) = ce^{-\left|\frac{k_1 x}{u}\left|\frac{2}{\alpha_E}\left(\sqrt{\alpha_E+1}-1\right)\right|\right|} \tag{6.54}$$

In the aforementioned formula, c is the integral constant and α_E is the dimensionless longitudinal dispersion coefficient, which is defined as follows: $\alpha_E = \frac{4E_x k_1}{u^2}$.

When $\alpha_E \to 0$ or $\sqrt{\alpha_E+1} \to 1$, the dispersion effect can be negligible, Eqns. (6.54) and (6.46) have the same solution of BOD distribution in the receiving river.

Considering that the point source pollution is discharged into the river at the mass rate of M BOD per unit time, and the river flow downstream of the discharge point $(x = 0)$ is Q. Because the BOD in the river is in a steady state, the BOD content discharged should be equal to the consumption of BOD in the river per unit time. According to the principle of mass conservation:

$$\int_{x_m}^{\infty} k_1 L(x)A\,dx = \dot{M}(x_m) \tag{6.55}$$

where x_m is the flow distance required for the point pollution source to achieve complete lateral mixing in the river and $\dot{M}(x_m)$ is the mass flux rate of the pollutant in the section x_m. Theoretically, $\dot{M}(x_m)$ is slightly smaller than $\dot{M}$.

Substituting Eqns. (6.54) into (6.55) and integrating the resulting equation by assuming $\dot{M}(x_m) \approx \dot{M}$, we have:

$$L_0 = \frac{\dot{M}}{Q}\left|\frac{2}{\alpha_E}\left(\sqrt{\alpha_E+1}-1\right)\right| \tag{6.56}$$

Example 6.6 CBOD of a river water shows a first-order deoxygenation reaction, and the deoxygenation coefficient is k_1.

Problem: (1) Calculate the required travel distance (x_D) when the initial concentration decays to the initial concentration multiplied by e^{-1}.

(2) Calculate the travel distance it takes for the discharged wastewater to become uniformly distributed laterally, x_m.

(3) Discuss the rationality of the one-dimensional steady-state advection–dispersion river water quality model.

Solution: (1) Calculate the required travel distance (x_D) when the initial concentration decays to the initial concentration multiplied by e^{-1}.

The first-order of CBOD deoxygenation is expressed as follows:

$$\frac{dL}{dt} = -k_1 L$$

When the initial condition is: $t = 0$, $L = L_0$, the solution of the aforementioned equation is expressed as follows:

$$L(t) = L_0 \exp(-k_1 t)$$

When the mixed sewage flows downstream, its travel time $\theta_t = \dfrac{x}{u}$ is the reaction time when the CBOD in the mixture is reduced to the initial concentration multiplied by e^{-1}.

$$\frac{L}{L_0} = \exp(-k_1 \theta_t) = \exp\left(-k_1 \frac{x_D}{u}\right) = e^{-1}$$

Therefore, the travel distance is expressed as follows:

$$x_D = \frac{u}{k_1}$$

(2) Calculate the travel distance it takes for the discharged wastewater to become uniformly distributed laterally, x_m.

According to the discussion in Section 6.1.2, if the sewage is discharged on the bank of the river, the distance x_m can be obtained by Eqn. (6.3a):

$$x_m = 0.4 \frac{u W^2}{\varepsilon_t}$$

(3) Applicability of the one-dimensional steady-state advection–dispersion river water quality model.

First, two distances are compared: one is the flow distance that the initial BOD concentration is roughly decayed, x_D, and the other is the flow distance required to achieve complete lateral mixing, x_m.

The ratio of the two distances is expressed as follows:

$$\frac{x_D}{x_m} = \frac{\dfrac{u}{k_1}}{0.4\dfrac{uW^2}{\varepsilon_t}} = 2.5\frac{\varepsilon_t}{k_1 W^2} \tag{6.57}$$

In Section 6.1.2, it is pointed out that in the absence of measured data, the lateral mixing coefficient and the longitudinal dispersion coefficient can be estimated by Eqns. (6.2) and (6.9):

$$\varepsilon_t = 0.6\,hu_*$$

$$E_x = 0.01\frac{u^2 W^2}{hu_*}$$

Combining the aforementioned two equations and eliminating hu_*, we can express dispersion coefficient E_x as a function of ε_t:

$$E_x = 0.006\,\frac{u^2 W^2}{\varepsilon_t} \tag{6.58}$$

Using Eqns. (6.57) and (6.58), the dimensionless longitudinal dispersion coefficient α_E can be written as follows:

$$\alpha_E = \frac{4E_x k_1}{u^2} = 0.024\frac{k_1 W^2}{\varepsilon_t} = 0.06\left(\frac{x_m}{x_D}\right) \tag{6.59}$$

Equation (6.59) implies:

(1) When the lateral mixing distance is less than the decay distance or $x_m < x_D$, the dimensionless longitudinal dispersion coefficient is less than 0.06 and can be ignored. Therefore, the Streeter–Phelps model, which does not consider dispersion, is still the most suitable model.
(2) When the lateral mixing distance is larger than the decay distance or $x_m > x_D$, the majority of oxygen-consuming organic carbon compounds has been decayed before the lateral mixing is completed. In this case, the one-dimensional river steady state model is not applicable, and the river water quality must be analyzed using a two-dimensional or three-dimensional model.

Comments

The Streeter–Phelps model assumes that the hydrodynamic transport mechanism of contaminants in the receiving river is advection dominant, and the dispersion can be ignored. In order to extend the Streeter–Phelps model, the first approach is to put the dispersion term back into the governing equation. However, this method is not suitable to simulate the transport of nonconservative pollutants according to the aforementioned example, as the longitudinal dispersion is either negligible or the 1D assumption is not valid.

The aforementioned example is for a nontidal river reach that takes a long time to complete lateral mixing. In a tidal river reach or in an estuary, the one-dimensional river steady-state water quality model with dispersion term can still be applied for a preliminary water quality model analysis.

6.3.2.2 Enhancement of reaction simulation

The traditional Streeter–Phelps model assumes that the natural processes affecting the DO include only CBOD deoxygenation and atmospheric reaeration (Figure 6.8). In reality, nitrification, benthic uptake, and plant photosynthesis/respiration also affect the DO in the receiving river. Considering these reaction mechanisms, the governing equation is Eqn. (6.36).

It is assumed that the photosynthetic/respiration term $[P(t) - R]$ of the plant in the aforementioned equation can be replaced by its daily average value (S_{P-R}), and Eqn. (6.36) becomes the following ordinary differential equation:

$$u\frac{dD(x)}{dx} = k_c L_c(x) + k_n L_n(x) - k_2 D(x) + S_B - S_{P-R} \tag{6.60}$$

There are three unknowns in Eqn. (6.60). CBOD L_c and NBOD L_n change with distance in the downstream reach. They can be calculated by the following simple first-order decay equation, and then, the result is substituted into Eqn. (6.59) to solve the change of DO:

$$u\frac{dL_c(x)}{dx} = -k_c L_c(x) \tag{6.61}$$

$$u\frac{dL_n(x)}{dx} = -k_n L_n(x) \tag{6.62}$$

A linear systems river BOD-DO model based on the modified Streeter–Phelps model consisting of the above three equations has been developed (Liu, 1996). The theoretical basis and application examples for this model are discussed in Section 6.5.

6.4 COMPREHENSIVE NUMERICAL RIVER WATER QUALITY MODELS

6.4.1 QUAL-2K model

QUAL-2K is the latest model in QUAL series for water quality simulation. The model is able to simulate various substances simultaneously, such as

BOD, DO, phosphorus, and nitrogen. Therefore, the model has been widely used for water quality assessment by researchers all over the world (Guo and Wang, 2003; Fang et al., 2008; Oliveira et al., 2012; Zhang et al., 2012; Sharma et al., 2017).

The governing equation and numerical methods of the QUAL-2K model have been introduced in Section 3.4. In this section, an example is presented to illustrate the application of QUAL2K model in water quality assessment. The example is based on the work by Fang et al. (2008), who have used the QUAL-2K model to predict the BOD and DO distribution in the Changshangang-Qujiang River and the Jinhua River. These two rivers are the main tributaries of Qiantang River, a sub-watershed of the Yangzi River, in China, as shown in Figure 6.10. The Changshangang-Qujiang River originates from the mountain area, whereas the Jinhua River flows through the region with more intense economic development.

In the model, the Changshangang-Qujiang River (146 km) and the Jinhua River (110 km) were considered as two subsystems, in which the Changshangang-Qujiang River was divided into 15 reaches and the Jinhua

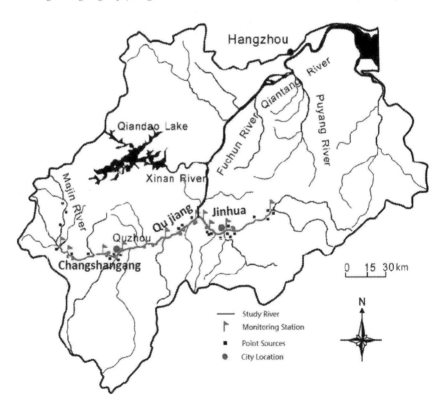

Figure 6.10 Qiantang River and its watershed.

Source: Fang et al., 2008

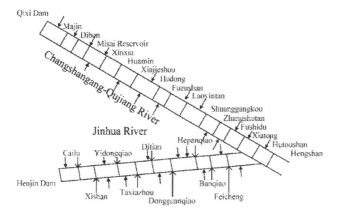

Figure 6.11 System segmentation and location of tributaries and pollution sources.
Source: Fang et al., 2008

River was divided into 10 reaches, as shown in Figure 6.11. Each reach was further divided into several uniform computational elements and then was solved using the QUAL-2K model.

In the model, the low flow rate (30Q 10 flow, a low flow of 30 days duration with recurrence interval of 10 years) was considered as the input condition. The simulated flow rate, BOD, and DO as well as their comparisons with the averaged field data along the two rivers are presented in Figure 6.12. The simulated results for both rivers showed good agreement with the field measurement. The BOD concentration in Changshangang-Qujiang River increased significantly at the distance of 60 km due to the waste water discharge from chemical plants and papermaking factories. The BOD concentration decreased in the downstream reaches due to low level of waste discharge. For the Jinhua River, on the other hand, the WWTPs did not run effectively so that the waste discharge increased along the river, resulting in an increase in the BOD concentration from upstream to downstream.

To carry out field measurement is always expensive and the measured conditions are also limited. To save time and money, the numerical model is a good prediction model, and its results can be used as a basis for decision making in practical project. In this study, QUAL-2K model, after validation, is further used to analyze the pollutant fate in the rivers under different scenarios. Scenario I was reducing 25% of the wastewater quantity from each plant and Scenario II was setting the BOD source levels at 30 mg/L. Figure 6.13 presents the simulated BOD and DO of the two rivers under two different scenarios. The results indicated that both two measures can reduce the BOD concentration in the rivers and the DO in the rivers increased accordingly. The simulated results by the QUAL-2K model provided an

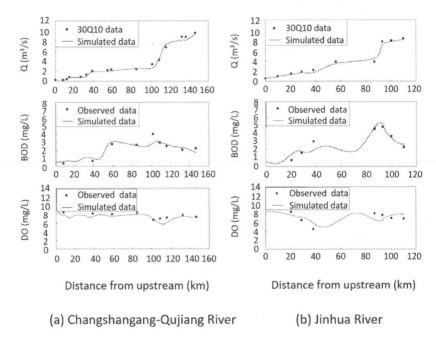

(a) Changshangang-Qujiang River   (b) Jinhua River

Figure 6.12 Comparisons of simulated results with the field data: (a) Changshangang-Qujiang River and (b) Jinhua River.

Source: Fang et al., 2008

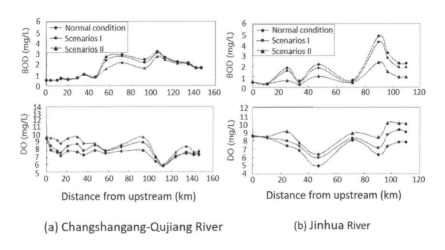

(a) Changshangang-Qujiang River   (b) Jinhua River

Figure 6.13 Simulated BOD and DO under different scenarios: (a) Changshangang-Qujiang River and (b) Jinhua River.

Source: Fang et al., 2008

intuitive results of water quality changes with the changes of discharge conditions. It can be concluded that better waste treatment and reducing wastewater discharge are helpful to improve the water quality.

6.4.2 WASP model

The WASP model is a numerical solution for river/lake water quality produced by the USEPA. The original WASP model was developed in the 1970s by an American engineering consultancy, Hydroscience, Inc. At that time, the company was commissioned by public and private organizations to conduct water quality survey of rivers, harbors, lakes, and estuaries. Successful examples include the analysis of the Great Lakes eutrophication (Hydroscience, 1973), the Hudson River Polychlorinated Biphenyl (PCB) Pollution Survey (Hydroscience, 1978), and the San Francisco Bay Water Quality Survey (Hydroscience, 1972).

As a mathematical simulation tool for water environment, one of the main features of WASP is its compartment-modeling program in different water bodies. WASP applies the FDM to solve the advection–diffusion reaction equation of the transport program and uses a forward-time and backward-space FD approximation to ensure numerical stability. See Section 3.2 or refer to the study by Smith (1965) for the principle of finite-difference numerical solutions.

WASP was structured for water quality modeling rather than hydrodynamics simulation. In the early WASP model analysis, the hydrodynamic equation for simulation of the water body was first established and solved, and then, the calculation results were used as the input data for pollutant transport modeling. The new version of WASP, WASP6, can be used in conjunction with other hydrodynamic programs to simulate hydrodynamics and pollutant transport simultaneously. Recently, Taipei University of Science and Technology combined the WASP model with the hydrodynamic model HEC-RAS, which will be introduced in Section 6.4.3, in the analysis of the water quality of the river.

6.4.3 HEC-RAS model

The HEC model is a river hydrology/hydraulic analysis model developed by the US Army Corps of Engineers Hydrologic Engineering Center in the 1960s. It has been widely used in flood forecasting and flood control engineering design for many years. The HEC-RAS increases the river water quality simulation function based on the original river hydrodynamic simulation (US Army Corps of Engineers, 2010). It then becomes an integrated model for multitasked water environment modeling, similar to WASP6.

The HEC-RAS model includes the following simulation modules: (1) steady flow and free surface level, (2) unsteady river flow, (3) sediment transport and bed morphology change, and (4) water quality. Each of these modules is interconnected by a common model network system and a hydrodynamic calculation program.

The river water quality program of the HEC-RAS model consists of a one-dimensional advection–diffusion reaction equation that simulates the transport of various pollutants and water quality parameters, including water temperature, DO, BOD, organic nitrogen, ammonia nitrogen, nitrate, nitrite, dissolved phosphorus, and algae.

It is noted that all of the models introduced earlier are for the analysis of 1D river flow and water quality. For lakes and reservoirs, sometimes it is necessary to model the 2D and even 3D flow and water quality. In the future, a new river water quality model that couples 2D or 3D flow and water quality should be developed.

6.5 LINEAR SYSTEM APPROACH TO RIVER WATER QUALITY MODELING

Fundamentals of system modeling, including theoretical basis and general application, are discussed in Section 2.5. Linear system modeling of a time-variable river system was presented as a chapter of an *American Society of Civil Engineers* (ASCE) book on 'Environmental Fluid Mechanics' (Liu and Neill, 2002).

In this section, the linear system modeling of a steady-state river system is presented. Figure 6.14 shows the changes of DO in a steady-state river

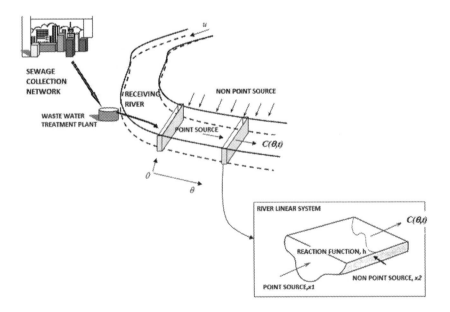

Figure 6.14 Schematic diagram of water quality model of river linear system.

receiving organic waste inputs. In this case, the governing equation takes the following form:

$$D(\theta_t) = D_s(\theta_t) + \int_0^{\theta_t} X(\tau) h(\theta - \tau) d\tau \tag{6.63}$$

where $D_s(\theta_t)$ is zero-input response function.

Successful application of a system model depends largely on the correct estimation of the system impulse response function, also known as system identification. System identification can be accomplished by using one of the following three methods: the method of physical parameterization, the method of system parameterization, and the method of system inverse (Liu, 1988). The detailed discussion of these three methods is presented in Section 2.5.3. The method of physical parametrization is used here to estimate the system response function of a linear steady-state river system. The other two methods are applied for watershed system modeling in Chapter 5 and soil transport system modeling in Chapter 8.

The physical parameterization method of system identification is to express the system impulse reaction function in terms of the physical parameters. Therefore, to apply this method, the physically based model of a steady-state river system must be derived first. The governing equation of the physically based model is a differential equation. When a δ function is used as the system's input function, the solution of this differential equation is the system impulse response function (see Section 2.5.2).

The physically based one-dimensional nondispersive steady-state river CBOD-DO model or the Streeter–Phelps model has the following governing equation:

$$\frac{dD(\theta_t)}{d\theta} + k_2 D(\theta_t) = X(\theta_t) \tag{6.64}$$

As shown in Figure 6.14, a linear river system is defined as a waste plume moving downstream with a time-of-travel θ_t, if waste loads entering the river at $\theta_t = 0$ consist of only carbonaceous wastes, the input function $X_1(\theta_t)$ to the system when the waste plume reaches a downstream location θ_t is expressed as follows:

$$X_1(\theta_t) = k_c L_{c0} \exp(-k_c \theta_t) \tag{6.65}$$

where X_1 is point source waste loading at θ_t; L_{c0} is CBOD loading at $\theta_t = 0$.

Figure 6.14 shows that the receiving river receives nonpoint source pollution load as well as point source pollution load. Nonpoint sources consist of the riverside landfill seepage and the benthic uptake. Therefore, when the waste plume reaches a downstream location θ, the nonpoint source input function X_2 can be written as follows:

$$X_2(\theta_t) = \int_0^{\theta_t} k_c L_{Sc} \exp(-k_c \xi) d\xi + L_{SB} \tag{6.66}$$

where $\int\limits_0^{\theta} k_d L_{Sc} \exp(-k_d \xi) d\xi$ is the nonpoint source pollution caused by the landfill seepage, L_{Sc} is the CBOD in the seepage water, and L_{SB} is the nonpoint source CBOD released from the river bottom.

With both point and nonpoint sources CBOD loadings, the input function X in a river linear system model is expressed as follows:

$$X(\theta_t) = X_1(\theta_t) + X_2(\theta_t) = k_c L_{C0} \exp(-k_c \theta_t)$$

$$+ \int_0^{\theta_t} k_c L_{Sc} \exp(-k_c \xi) d\xi + L_{SB} \tag{6.67}$$

When the river system receives a pulse input of $L_{Sc} = \delta(\theta_t)$, the system input function $X_1(\theta_t)$ in Eqn. (6.64) is $X_1 = k_c \delta(\theta_t)$. Recall that the analytical solution to Eqn. (6.67) receiving a pulse input is the system response function $h(\theta_t)$ or:

$$h(\theta_t) = k_c \exp(-k_2 \theta_t) \tag{6.68}$$

where $h(\theta_t)$ is the system impulse response function of a linear river system as illustrated by Figure 6.15.

Note that the system response function of a river linear system model represents the river's self-purification ability. In this case, the time of travel θ_t and the reaction coefficients k_c and k_2 represent the influence of the hydrodynamic mechanism and the natural reaction mechanisms of the river's self-purification ability.

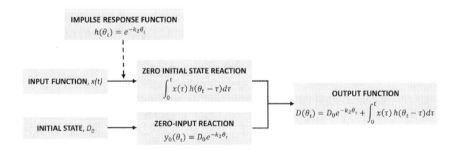

Figure 6.15 BOD-DO model analysis of a steady-state river linear system.

The zero-input reaction in the linear river system model Eqn. (6.66) or D_s (θ_t) is due to the initial state of the system, or:

$$D_s(\theta) = D_0 \exp(-k_2\theta_t) \tag{6.69}$$

Substituting Eqns. (6.67)–(6.69) into Eqn. (6.63), the linear system BOD-DO model of a receiving river as shown in Figure 6.15 is expressed as follows:

$$D(\theta_t) = D_0 \exp(-k_2\theta_t)$$
$$+ \int_0^{\theta_t} \left[k_c L_{c0} \exp(-k_c\theta_t) + \int_0^{\theta_t} k_c L_{Sc} \exp(-k_c\xi) d\xi + L_{SB} \right] k_c \exp(-k_2\theta_t) d\tau \tag{6.70}$$

Example 6.7 Linear river system modeling.

Problem: A river is subjected to wastewater effluent from a municipal waste water treatment plants and seepage from an area landfill (Figure 6.14). The total flow after mixing the dry flow and the discharge water is 5.0 m/s. The CBOD concentration is 20.0 g/m³, the DO is 6.9 g/m³, the water temperature is 24°C, and the DO saturation is 8.4 g/m³. The load of nonpoint source pollution caused by sewage infiltration of the landfill is 15.0 kg/m³/km. The flow rate is small, and its impact on the total flow of the river can be neglected. According to the water quality measurement of the river, the oxygen consumption coefficient and the reaeration coefficient were 0.23 d⁻¹ and 0.46 d⁻¹, respectively.

Apply the linear BOD-DO model to analyze and evaluate various schemes to improve the river self-purification capacity and reduce pollution load, including:

(1) Increase the reaeration coefficient from 0.46 to 0.92 d⁻¹.
(2) Reduce the point source pollution load.
(3) Reduce the nonpoint source pollution load and completely eliminate the sewage infiltration in the landfill.

Solution: The impulse response function of the DO model of the river linear system is expressed as follows:

$$h(\theta_t) = \exp(-k_2\theta_t) = \exp(-0.46\theta_t)$$

The zero-input response function of the DO model of the river linear system is given as follows:

$$D_s(\theta_t) = D_0 \exp(-k_2\theta_t) = (8.4 - 5.9)\exp(-0.46\theta_t) = 2.5\exp(0.46\theta_t)$$

The point source input function X_1 of the river linear BOD-DO model is given as follows:

$$X_1(\tau) = k_c L_{c0} \exp - k_c \tau = (0.23)(20)\exp(-0.46\tau)$$

The input function X_2 of nonpoint source caused by sewage infiltration of the landfill is expressed as follows:

$$X_2(\tau) = \int_0^\tau k_c L_{Sc} e^{-k_c\xi} d\xi$$

$$= \int_0^\tau 0.23(15)\exp(-0.23\xi)d\xi = 2.6[1 - \exp(-0.23\tau)]$$

By substituting the input function and impulse response function into Eqn. (6.67), the function of DO deficiency in the river is expressed as follows:

$$D(\theta_t) = 2.5\exp(-0.46\theta_t)$$

$$+ \int_0^{\theta_t} [2.0\exp(-0.23\tau) + 2.6]\exp[-0.46(\theta_t - \tau)]d\tau$$

As $DO(\theta_t) = C_s - D(\theta_t)$, we have:

$$DO(\theta_t) = 8.4 - \left\{ 2.5\exp(-0.46\theta_t) + \int_0^{\theta_t} [2.0\exp - (0.23\tau) \right.$$

$$\left. + 2.6]\exp[-0.46(\theta_t - \tau)]d\tau \right\}$$

(1) Figure 6.16 shows the effect of reducing CBOD on the DO. When the CBOD of the discharge point is 20.0 g/m³, the minimum DO in the receiving river is only about 1.0 g/m³. When the CBOD is reduced to 10 g/m³, the minimum DO is increased to more than 2.0 g/m³.

(2) Figure 6.17 shows the effect of increasing reaeration coefficient from 0.46 to 0.92 d⁻¹ on the DO. The reaeration coefficient of this reach is 0.46 d⁻¹, which is measured by the tracer method. When the measured value of reaeration coefficient is used, the lowest DO in the receiving river is only about 1.0 g/m³. The reaeration capacity of rivers can be improved by the change of their hydraulic properties. If the reaeration coefficient increases to 0.92 d⁻¹, the minimum DO will increase to more than 4.0 g/m³.

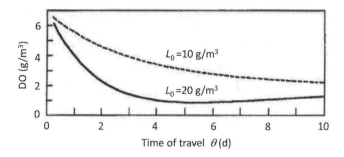

Figure 6.16 River DO content variation by reducing its BOD loading from 20 to 10 g/m³.

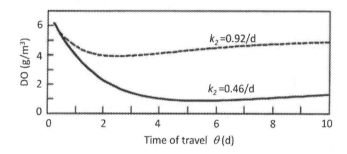

Figure 6.17 River DO content variation by increasing its reaeration coefficient from 0.46 to 0.92 m⁻¹.

Comments

The water quality of receiving river depends on its self-purification capacity and pollution load. For the aerobic organic waste, the self-purification capacity of the river is determined by the reaeration coefficient. This example points out that when the coefficient of reaeration is doubled, the minimum content of the DO is increased fourfold. Therefore, it is an economically efficient action for water pollution control to improve the capacity of reaeration by changing the hydraulic properties of the river.

EXERCISES

1. Example 6.7 uses the linear river BOD-DO system to analyze the efficiency of various water pollution control measures. If all conditions in Example 6.7 are unchanged but ignore the sewage infiltration of the landfill, calculate the minimum DO of the river.
2. A design project of linear river system modeling.

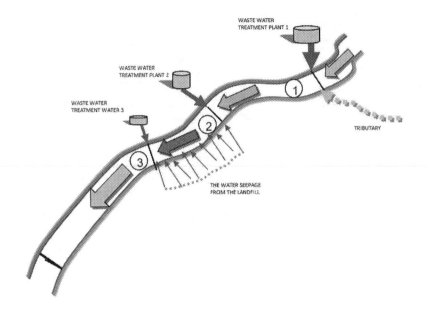

Figure 6.18 Division of the Qingshui River in the modeling analysis.

Background: An environmental engineering company has signed a contract with the city of Holton for a sewage treatment plant improvement project. At present, there are three sewage treatment plants in Horton city to discharge the treated sewage into the Qingshui River, and the DO of the Qingshui River is lower than the required water quality standard of 4.0 mg/L.

According to the contract, the engineering company needs to analyze the river water quality characteristics and draw up the planning report. Various economic and effective measures including reducing pollutant discharge to meet the water quality standard of the Qingshui River should be proposed. Because you have taken the course of the water quality modeling analysis and have experience in the application of linear system model for river network, you are appointed as the chief engineer of this planning project by the company.

Relevant Data: In order to apply the linear system model of river network, the Qingshui River is divided into three reaches. The discharge water from the first sewage treatment plant and a tributary flow into the Qingshui River at the beginning of the first reach. The discharge water from the second and third sewage treatment plants flows into the Qingshui River at the beginning and end of the second reach, respectively (Figure 6.18).

In collaboration with the engineering consultants, Horton city conducted intensive water quality measurement in the Qingshui River to collect data related to its self-purification capacity and pollution load. The data are shown in Tables 6.7–6.9. In addition, there is a landfill on the right bank of the third reach. The water seepage from the landfill discharges 1,250 pounds of CBOD and 1700 pounds of NBOD into the Qingshui River per day per mile in the form of regular nonpoint source pollution. Because the discharge of landfill seepage is very small, its influence on river flow and other hydraulic parameters can be ignored. The reaeration coefficient of each reach can be estimated by O'Connor and Dobbins' formula (Table 6.5).

Table 6.7 Hydraulic and water quality parameters of each reach and tributary of the receiving river

Parameters	Esturay	Reach 1	Reach 2	Reach 3	Tributary
Reach length (miles)	–	6.0	5.0	10.0	–
Starting height of the reach (ft)	–	685	670	550	–
Upstream flow (cfs)	10.0	–	–	–	0.25
Velocity (fps)	–	0.29	0.43	0.22	–
Water depth (ft)	–	1.0	1.0	1.5	–
Temperature (°C)	–	23.0	21.0	25.0	18.0
Dissolved oxygen deficit (D, g/m³)	2.3	–	–	–	3.3
Oxycarbide (CBOD, g/m³)	2.0	–	–	–	1.5
Oxynitride (NBOD, g/m³)	2.0	–	–	–	1.5

Table 6.8 Hydraulic and water quality parameters of discharged water from the sewage treatment plants

Parameters	Waste Water Treatment Plant 1	Waste Water Treatment Plant 2	Waste Water Treatment Plant 3
Flow (MGD)	6.0	8.0	1.1
Oxycarbide (CBOD, g/m³)	150	105	30
Oxynitride(NBOD, g/m³)	60	30	35
Dissolved oxygen deficiency (D, g/m³)	6.3	6.3	6.3

Table 6.9 Estimated values of model coefficients

Coefficient	Reach 1	Reach 2	Reach 3
k_d, d⁻¹	0.60	0.60	0.44
k_r, d⁻¹	1.24	1.24	1.00
k_n, d⁻¹	0.50	1.30	1.30

Design Project Requirements: The design project includes three parts: model derivation, water quality analysis of the Qingshui River, and management scheme evaluation. The detailed requirements are as follows:

(1) Derivation of DO linear system model

The following pollution loads should be considered in the DO linear system model of the Qingshui River:

    (a) Point source DO deficit
    (b) Point source CBOD
    (c) Point source NBOD
    (d) Nonpoint source CBOD
    (e) Nonpoint source NBOD
    (f) Nonpoint source benthal demand effect

In the model analysis, the Qingshui River is divided into reaches with similar hydraulic and natural response properties. At the beginning of each reach, the mass conservation analysis is carried out to obtain the initial conditions of the reach. The analysis shall take into account the flow and water quality from upstream, tributaries, and sewage treatment plants. Relevant calculation programs can be written and executed by Excel or other computer software.

(2) Use the model to simulate the current water quality of the Qingshui River.

The water quality parameters calculated by the model include the change in oxygen-consuming carbide (CBOD), oxygen-consuming nitride (CBOD), and DO along the Qingshui River. Draw the concentration curve of each parameter. Based on the calculation results, the water quality and pollution problems of Qingshui River are discussed.

(3) Analyze and evaluate various feasible water quality management schemes by using the model analysis:

    (a) Calculate the removal rate of CBOD and NBOD in the discharge water of each sewage treatment plant to meet the DO standard of the Qingshui River.
    (b) The feasibility of reducing the seepage water from the landfill to reach the DO standard of Qingshui River is evaluated by using the modeling analysis.
    (c) The feasibility of improving the reaeration efficiency of the Qingshui River to meet the DO standard is evaluated by using the modeling

analysis. One possible way to improve the reaeration efficiency is to build a barrage at the beginning of the second reach to form an artificial waterfall with a water level drop of 3 m. This improvement on DO can be expressed by the following formula:

$$\gamma_D = 1 + 0.38 f_a f_b h'(1 - 0.11h')(1 + 0.046T) \tag{6.71}$$

where γ_D is the ratio of DO deficiency in the upstream and downstream of the artificial waterfall, h' is the fall of the artificial waterfall (meters), T is the water temperature (°C), the design standard water temperature is 20°C, f_a is the empirical constant related to the water quality (f_a = 0.65–1.05 in moderately polluted river), and f_b is the empirical constant related to the shape and quality of the barrage (f_b = 0.75 in circular curve type barrage).

(4) Write design project report.

The report should discuss the results of modeling analysis, propose a water quality management plan, and discuss the feasibility and future development of the linear system river water quality model.

REFERENCES

Churchill, M.A., Elmore, H.L. and Buckingham, R.A. (1962). The prediction of stream reaeration rates. *Journal of the Sanitary Engineering Division*, 88(4).

Fang, X., Zhang, J., Chen, Y. and Xu, X. (2008). QUAL2K model used in the water quality assessment of Qiantang River, China. *Water Environment Research*, 80(11), pp. 2125–2133.

Fischer, H.B., List, E.J., Imberger, J. and Brooks, N.H. (1979). *Mixing in Inland and Coastal Waters*. Academic Press, New York.

Guo, Y.B. and Wang, Y.X. (2003). Hydrochemical modeling and prediction of the middle and lower reaches of Hanjiang river: Application of QUAL2K model. *Safety and Environmental Engineering*, 10(1), pp. 4–7 (in Chinese).

Hydroscience, Inc. (1972). *Mathematical model of the phytoplankton dynamics in the Savremento-San Joaquin Bay Delta*. Preliminary Report, Prepared for the California Department of Water Resources.

Hydroscience, Inc. (1973). *Limnological system analysis of the agreat Lakes: Phase I – Preliminary model design*. Hydroscience, Inc., Mahwah, NJ.

Hydroscience, Inc. (1978). Estimation of PCB reduction by remedial action on the Hudson River. *Prepared for the State of New York*, Department of Environmental Conservation.

Liu, C.C.K. (1986). Surface water quality analysis. Chapter 1 in Wang, L. and Pereira, N.C. (eds.), *Handbook of Environmental Engineering*. The Humana Press, Totowa, New Jersey, pp. 1–59.

Liu, C.C.K. (1988). Solute transport modeling in heterogeneous soils: Conjunctive application of physically based and system approaches. *Journal of Contaminant Hydrology*, 3(1), pp. 97–111.

Liu, C.C.K. (1996). Linear systems theory and modeling of river water quality. *Journal of the Chinese Institute of Environmental Engineering* (Taipei), 6(1), pp. 51–58.

Liu, C.C.K. and Guo, F. (1994). Time series analysis of initial dilution at the Sand Island Outfall, Oahu, Hawaii. *The Sixth Asian Congress of Fluid Mechanics, May 22–26, 1995, Singapore*, pp. 1142–1145. Available at: https://www.tib.eu/en/search/id/TIBKAT%3A216703514/ Proceedings-of-the-Sixth-Asian-Congress-of-Fluid/.

Liu, C.C.K. and Neill, J.J. (2002). Chapter 12. Linear systems approach to river water quality analysis. In *Environmental Fluid Mechanics – Theories and Application*. ASCE Book, American Society of Civil Engineers, Reston, pp. 421–457.

O'Connor, D.J. and Di Toro, D.M. (1970). Photosynthesis and oxygen balance in streams. *Journal of the Sanitary Engineering Division*, 96(2), pp. 547–571.

O'Connor, D.J. and Dobbins, W.E. (1958). Mechanism of reaeration in natural streams. *American Society of Civil Engineers*, 123, pp. 641–684.

Oliveira, B., Bola, J., Quinteiro, P., Nadais, H. and Arroja, L. (2012). Application of Qual2Kw model as a tool for water quality management: Cértima river as a case study. *Environmental Monitoring and Assessment*, 184(10), pp. 6197–6210.

Owens, M., Edwards, R.W. and Gibbs, J.W. (1964). Some reaeration studies in streams. *International Journal of Environment and Pollution*, 8, pp. 469–486.

Qin, B., Zhu, G., Gao, G., Zhang, Y., Li, W., Paerl, H.W. and Carmichael, W.W. (2010). A drinking water crisis in Lake Taihu, China: Lingake to climate variability and lake management. *Environmental Management*, 45, pp. 105–112.

Sharma, D., Kansal, A. and Pelletier, G. (2017). Water quality modeling for urban reach of Yamunariver, India (1999–2009), using QUAL2Kw. *Applied Water Science*, 7(3), pp. 1535–1559.

Smith, G.D. (1965). *Numerical Solution of Partial Differential Equations*. Oxford University Press, London.

Sokolnikoff, I.S. and Redheffer, R.M. (1966). *Mathematics of Physics and Modern Engineering*. McGraw-Hill Book Company, New York.

Taylor, G.I. (1954). The dispersion of matter in turbulent flow through a pipe. Proceedings of the Royal Society of London. Series A. *Mathematical and Physical Sciences*, 223(1155), pp. 446–468.

Thomann, R.V. (1972). *Systems Analysis and Water Quality Management*. McGraw-Hill, New York.

Tsivoglou, E.C. and Wallace, J.R. (1972). *Characterization of Stream Reaeration Capacity*. U.S. Environmental Protection Agency Ecological Research Series, Report EPA-R3-72-012, 317 P.

U.S. Army Corps of Engineers. (2010). *HEC-ARS River Analysis System Hydraulic Reference Manual*. Version 4.0, CPD-69, U.S. Army Corps of Engineers Hydraologic Engineering Center.

Vesilind, P.A. and Peirce, J.J. (2013). *Environmental Engineering*. Ann Arbor Science the Butterworths Group.

Zhang, R., Qian, X., Yuan, X., Ye, R., Xia, B. and Wang, Y. (2012). Simulation of water environmental capacity and pollution load reduction using QUAL2K for water environmental management. *International Journal of Environmental Research and Public Health*, 9(12), pp. 4504–4521.

Zison, S.W., Mills, W.B., Deimer, D. and Chen, C.W. (1978). *Rates, Constants, and Kinetics Formulations in Surface Water Quality Modeling*. U.S. Environmental Protection Agency Research Report Series, EPA-600/3-78-105.

Chapter 7

Intensive river survey in river water quality modeling

7.1 PLANNING AND EXECUTION OF AN INTENSIVE RIVER WATER QUALITY SURVEY

An intensive river water quality survey is conducted to collect field data that can be used to evaluate the model parameters characterizing the river's hydraulics and reaction kinetics. In this chapter, for simplicity, creeks, streams, and rivers are referred to as rivers.

Without intensive water survey data, the values of model parameters of a water quality river model must be determined on the basis of published literatures or by using data collected from laboratory experiments. The values in published literature are often derived under different river conditions than the modeled river, and laboratory experiments have limited abilities to simulate all of the field conditions that affect the values of model parameters. Ideally, river water quality modeling is conducted with model parameters determined directly from a carefully planned and executed intensive field survey. This chapter introduces the procedures and techniques of the planning and execution of intensive river water quality surveys.

An intensive river water quality survey is different from a river water quality monitoring program. A water quality monitoring program of a particular river is conducted to collect data at fixed river locations, and thus, monitoring data can only be used to delineate the state of river water quality and its historical trend. Conversely, an intensive river water quality survey is conducted by tracking the waste plumes when they are moving downstream, and thus, data collected in an intensive water quality survey of a particular river provide the cause–consequence relationships between waste loadings and river water quality and can be used to determine values of model parameters more reliably.

Municipal and industrial WWTPs discharge their effluents into receiving rivers by individual outfalls. These effluents are considered as point source waste loadings. In the river water quality management, point source waste loadings are regulated through mandatory discharge permits, such that the receiving river water quality standards are not contravened, even under the most critical

DOI: 10.1201/9781003008491-7

conditions (see Chapter 1). For most of the receiving rivers, critical conditions occur during low flow periods. Therefore, river intensive water quality surveys are generally conducted during low flow periods when the river flow is relatively steady such that the model parameters determined on the basis of the data collected in these surveys reflect the intended modeling conditions.

7.1.1 The procedure of conducting an intensive river water quality survey

A comprehensive intensive river water quality survey is conducted in six tasks (Figure 7.1).

7.1.1.1 Preliminary survey plan

A preliminary survey plan is prepared to estimate the river length and reaches to be included in the survey, locations of sample collecting stations, and the number of samples of each relevant substance/contaminant to be collected and the sampling frequency. This task can be conducted by collecting and reviewing all readily available information on the sources of wastes, water uses, and river characteristics. Generally, a quick search of the websites of governmental agencies, universities, and professional societies can yield most of the needed information.

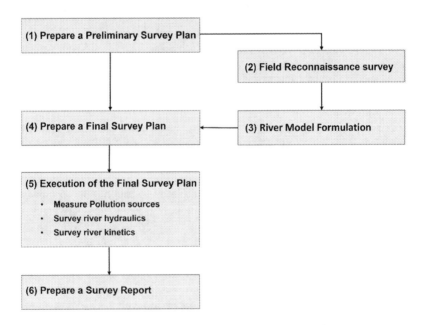

Figure 7.1 The procedure of conducting an intensive river water quality survey.

The lengths of river reaches can be determined on the basis of the estimated reaction rate and time of travel of the principal transporting substances or contaminants. The locations of sampling stations can be selected on a map by considering the accessibility. Sampling frequency can be selected by considering the contaminant reaction kinetics. For example, only one composite BOD sample needs to be collected at each sampling station and is then sent to a laboratory for further long-term BOD analysis; DO must be measured in situ at regular intervals (usually every 2 h) throughout a day to allow the plotting of a diurnal DO variation curve.

7.1.1.2 Field reconnaissance survey

A field reconnaissance survey is conducted to collect additional information for the formulation or selection of a suitable river model and for the preparation of a final survey plan. The field reconnaissance team is led by the survey team leader who is normally an environmental engineer. The team also consists of a hydrologist, an environmental chemist, a biologist, an instrument technician, and other specialists. The team members should meet at the site for a briefing on the water uses and pollution sources before separating to conduct their works individually.

One of the major outputs of a reconnaissance survey is to estimate the arrival time of the pollutant plume at various downstream points in order to select sampling stations. Another major output is to identify potential sites for a mobile field laboratory – ideally, this field laboratory is located at an existing water or WWTPs.

7.1.1.3 River model formulation

The model formulation must be completed before the final survey plan because the principal reason for conducting an intensive river water quality survey is to produce field data for accurate determination of the values of river model parameters.

Based on the information collected during the preliminary field survey and relevant literature, a receiving river water quality model is formulated and used to determine the allowable waste loading in the river such that the water quality standards established by the regulatory agency can be met.

7.1.1.4 Final survey plan

A final survey plan is prepared by revising the preliminary plan based on information obtained from a field reconnaissance survey and the preliminary river water quality modeling results.

7.1.1.5 Execution of the final survey plan

After the completion of the final survey plan, the team leader calls for a team meeting to inform team members of the final plan and to assign responsibility. Before the field survey starts, Rhobamine WT dye or other tracers should be prepared and the fluorometers' checked and calibrated. Water sample collectors should familiarize themselves with traffic route between sampling stations and technical details of samplers and other instruments to be used. Generally, team members should be at the survey site a couple days before the survey starts.

After the survey starts, collected data should be analyzed and plotted on a daily basis as preliminary charts and diagrams. By doing so, any problems in the final plan can be identified and corrected.

During the field survey, the team leader or another designated team member should visit all WWTPs, which discharge their effluent into the survey reaches and should collect effluent samples and analyze them in the field laboratory.

During the survey, the team leader and another designated team member should also investigate areas adjacent to the river reaches to identify any other waste sources such as any untreated municipal and industrial wastewater inflow, runoff from farmland, and seepage from municipal solid waste landfills.

7.1.1.6 Survey report

All data collected in an intensive river water quality survey should be analyzed to estimate hydraulic and reaction kinetic coefficients of the survey river as soon as possible. The survey report, which includes all collected data and the estimated coefficients, is prepared and usually becomes an important part of the modeling report.

7.1.2 An example: intensive river water quality survey in Canandaiqua Outlet

The model formulation and intensive water quality survey in a past water quality management study of the Canandaiqua Outlet, a tributary to Genesse River in Upstate New York, is used to illustrate the conduct of an intensive field survey.

Before 1980, the Canandaiqua Outlet in Upstate New York, USA, was heavily polluted due to wastewater discharge from several WWTPs in the area. An intensive river water quality survey was conducted during a period of July 17–20, 1978, to assess the water quality condition and to collect data for the determination of a river water quality model.

A map was prepared to show the selected survey locations of river reaches, wastewater discharge points, locations of the selected sampling stations, and highways and bridges. The final survey map for the 1978 river intensive water quality survey in Canandaiqua Outlet is shown in Figure 7.2. In order to facilitate the deployment of sampling instruments and to reduce the time required to deliver the collected water samples to the laboratory, most of the sampling stations are at or near bridges.

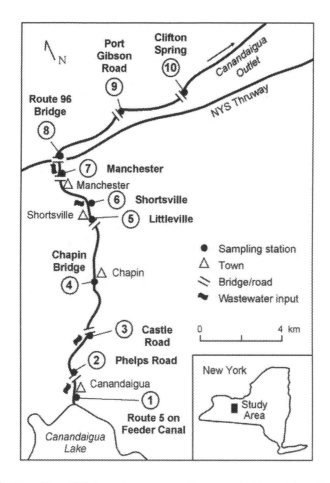

Figure 7.2 Map of the 1978 intensive water quality survey in Canandaiqua Outlet, New York, USA.

Table 7.1 The summary of the field survey of Canandaiqua Outlet, New York in 1978

Treatment Plant	Distance to the Mouth (km)	Discharge (m³/d)	DO (g/m³)	CBOD (g/m³)	NBOD (g/m³)
Canandaiqua	0.73	8900	5.0	65	125
VA Hospital	2.33	947	4.2	65	50
Shortsville	13.60	606	3.0	375	220
Manchester	16.90	909	3.0	580	179

Table 7.1 presents the location and effluent characteristics of four WWTPs along the survey reaches of Canandaiqua Outlet. It summaries the information of area WWTPs collected and analyzed in July 1978 survey, including the location of four WWTPs along the Canandaiqua Outlet and their effluent flow rate, DO content, and organic waste loadings in terms of CBOD and NBOD. The US Federal Clean Water Act requires all municipal treatment plants to provide a minimum secondary treatment. During the time of the survey, all four treatment plants could not meet this requirement. For example, CBOD concentrations in the effluents of these four treatment plants range d from 65 to 580 65 g/m³, which were much higher than the secondary effluent of 65 g/m³ or less.

On the basis of information collected during the preliminary field survey and by referring to the relevant literatures, a one-dimensional steady-state Canandaiqua Outlet model as shown below was formulated. This model can be used to determine if the minimum secondary treatment or higher level of treatment are needed at each of the four treatment plants to meet the water quality standards of Canandaiqua Outlet established by the New York State Department of Environmental Conservation under the Clean Water Act.

$$u\frac{dD}{dx} = E\frac{d^2D}{dx^2} + k_dL_c + k_nL_n - k_2D - (P_{DO} - R) \tag{7.1}$$

where P_{DO} is the photosynthetic oxygen production coefficient (mg/L/d) and R is the algal respiration coefficient (mg/L/d).

Equation (7.1) consists of seven model parameters. Among them, u and E are hydraulic parameters and k_1, k_n, k_2, S_B, P_{DO}, and R are reaction parameters.

Intensive river water quality surveys are often conducted for the accurate determination of the parameters of steady-state river water quality models (Liu, 1986). Nonpoint source pollution of a receiving water is caused by storm runoff and the contaminants it carries, and the river flow after a heavy storm is under unsteady-state conditions.

Therefore, river water quality modeling for nonpoint source pollution control must be able to simulate the river under unsteady-state flow conditions, which is much more complicated than steady-state modeling. An example of river water quality modeling under unsteady-state flow conditions and the conduct of field data for model calibration are discussed in Chapter 5.

7.2   RIVER INTENSIVE WATER QUALITY SURVEY OF HYDRAULIC PARAMETERS

7.2.1   River velocity and flow rate

The flow velocities of large rivers are traditionally measured by using a current meter. During a field survey, a river cross section is divided into a number of small area elements, and the velocity of each element is measured by a current meter. A multiplication of the velocity and area of each small element gives the flow rate of each element. The total river flow rate is then determined by summing up the values of individual elements (Turnipseed and Sauer, 2010).

The flow rate of small river or stream can be measured by using a weir or a Venture type flume, which are installed in a stream to create a critical flow condition such that a simple relation between flow and water depth exists (Li, 1983). The flow and water depth relation of a particular weir or a Venturi flume is calibrated at a hydraulic laboratory. The stream flow can then be determined readily by the measured stream water depth (Linsley et al., 1992). In recent years, modern devices for river flow and velocity measurements such as the Accoustic Doppler Current Profile (ADCP) have been developed and widely used (Mueller et al., 2009).

Figure 7.3 shows the three survey stations selected in a water quality survey conducted on April 9, 2013, along the Fucheng River in Chengdu, China. The velocity profiles at these three stations were measured by using an ADCP, from which the flow rate could be calculated. The velocity measurement at each station took only about 30 minutes. Figure 7.4 shows the measured digital data of the velocity contour at the station B-B. The data analysis software calculated the flow rate immediately based on the collected digital data. During the time of the survey, the river flow at station B-B was 7.4 m^3/s and the average cross-sectional velocity was 0.29 m/s.

7.2.2   Time of travel

Tracers are often used to measure the time of travel of contaminants in a river moving downstream. A tracer is a chemical that does not react in river

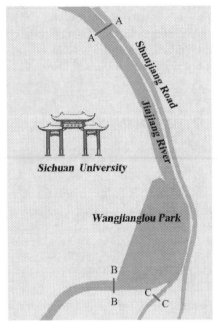

(b) ADCP flow/velocity measuring device

(c) Data analysis auxiliaries

(a) Location map

Figure 7.3 Measuring velocity profiles and flow rate in Fucheng River, Chengdu, China by an ADCP: (a) location map, (b) ADCP flow/velocity measuring device, and (c) data analysis auxiliaries.

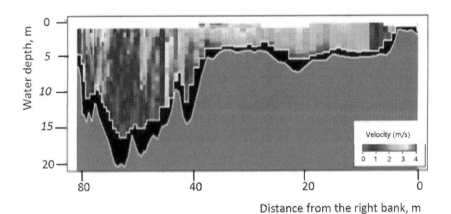

Figure 7.4 Measured velocity distribution at monitoring station B-B.

water and can be measured accurately in low concentration. In hydrologic and water quality surveys, US Geological Survey uses Rhodamine WT dye as a tracer, which is a conservative chemical and can be readily measured in a receiving stream by a fluorometer (Wilson, 1968).

Rhodamine WT dye may be released into a river in two different modes: (1) to release a large amount of the dye solution into a river in a short time and thus can be represented mathematically by a Dirac delta function; and (2) to release a small amount of the dye solution steadily into a river in a constant amount and thus can be represented mathematically by a step function (Rathbun et al., 1975).

Figure 7.5 shows a schematic diagram of dye tracer releasing into a stream as part of a river intensive water quality survey. Point A is the dye releasing point, B-B is the estimated downstream location where the complete vertical mixing is achieved, and C-C is the estimated downstream location where the complete cross-sectional mixing is achieved. Because only the river reaches below the point C can be taken as a one-dimensional system, the dye sampling stations for a one-dimensional analysis are located below the point C (see Section 6.1.1).

The measured tracer concentration curves $C(x_1,t)$ and $C(x_2,t)$ are plotted in Figure 7.6, where t_{p1} and t_{p2} are the times when the peak tracer concentration was observed at Stations 1 and 2, respectively. The time of travel between these two stations θ_{s1-s2} can be calculated as follows:

$$\theta_{s1-s2} = t_{p2} - t_{p1} \tag{7.2}$$

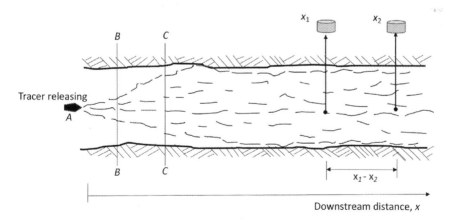

Figure 7.5 Measurement of the time of travel in a river using dye tracers.

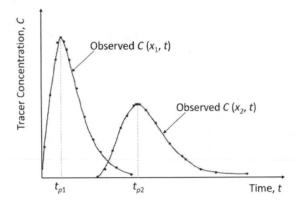

Figure 7.6 Tracer concentration curves to be used for estimating time of travel.

7.2.3   Longitudinal dispersion coefficient

The transport of a dye tracer plume in a one-dimensional river system can be simulated by the following equation:

$$\frac{\partial C}{\partial t} = -u\frac{\partial C}{\partial x} + E\frac{\partial^2 C}{\partial x^2} \qquad (7.3)$$

Note that the dye tracer is a conservation chemical, and there is no reaction terms in Eqn. (7.3).

If the dye tracer is released into the receiving stream as a pulse input, the initial/boundary conditions take the following form:

$$C(x,0) = M\delta(0) \qquad (7.4)$$

The analytical solution of this model is (see Section 2.2.3),

$$C(x,t) = \frac{M_r}{\sqrt{4\pi Et}} e^{-\left[\frac{(x-ut)^2}{4Et}\right]} \qquad (7.5)$$

where M_r is the mass of dye tracer released. If $M_r = 1$, $C(x,t)$ in Eqn. (7.5) is the impulse response of the one-dimensional system, or $h(x,t)$.

By taking the river reach between two sampling stations x_1 and x_2 as a linear system, the dye concentration curves observed at the upstream station $C(x_1,t)$ can be taken as the input function and the dye concentration curves at the downstream station $C(x_2,t)$ can be taken as the output function.

According to the linear system theory and its application for water environment modeling (see Chapter 2), with $C(x_1,t)$ as the input function, the output function can be calculated as follows:

$$C_\cdot\left(x_2,t\right) = \int_0^t C\left(x_1,\tau\right)\frac{M_r}{\sqrt{4\pi E\left(t-\tau\right)}} e^{-\frac{\left\{\left(x_2-x_1\right)-\left[u(t-\tau)\right]\right\}^2}{4E(t-\tau)}}\, d\tau \tag{7.6}$$

where $C_\cdot(x_2,t)$ is the output function or the calculated dye concentration curve at station x_2.

The shape of the downstream dye tracer concentration $C_\cdot(x_2,t)$, which is calculated by Eqn. (7.6) and shown in Figure 7.6, depends on the value of dispersion coefficient E. By using the LS method or the trial-and-error method, a particular value of E which gives the closest match of $C_\cdot(x_2,t)$ and $C(x_2,t)$ is the estimated E value of the river reach.

7.3   RIVER INTENSIVE WATER QUALITY SURVEY OF KINETICS PARAMETERS

7.3.1   Carbonaceous BOD deoxygenation coefficient

Deoxygenation coefficient of organic materials in terms of CBOD can be determined by batch experiment in a laboratory BOD bottle (see Section 6.2.1). However, stream environmental conditions that affect CBOD deoxygenation are difficult to reproduce satisfactorily in a laboratory. Ideally, river water quality modeling must be conducted by using the CBOD deoxygenation coefficient measured as part of a field intensive survey.

The time of travel of a waste plume in a river moving downstream is also the reaction time of CBOD in the plume. As discussed in Section 7.2.2, the time of travel of a waste plume moving downstream is measured as part of the intensive river water survey. As CBOD decomposes in a first-order reaction, by plotting measured BOD_5 in a waste plume in logarithmic scale along a receiving river with respect to corresponding time of travel would yield a straight line (Figure 7.7). The slope of this straight line is the CBOD deoxygenation coefficient.

Example 7.1 Measuring CBOD deoxygenation coefficient based on the following data collected by an intensive field water quality survey.

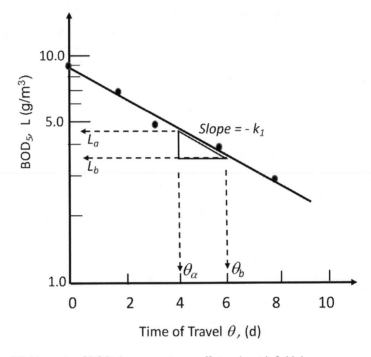

Figure 7.7 Measuring CBOD deoxygenation coefficient k_1 with field data.

Problem: Use values of BOD$_5$ of the water samples collected along a receiving stream as follows and determine the stream CBOD deoxygenation coefficient.

Survey Data	Station 1	Station 2	Station 3	Station 4	Station 5
Time of travel, θ (d)	0	1.9	3.2	5.5	7.8
BOD$_5$, L (g/m³)	9.0	6.9	5.0	3.4	2.7

Solution: Equation (6.18) was modified to the following form for determining the CBOD deoxygenation coefficient with field data:

$$k_c = -\frac{\log L_b - \log L_a}{\theta_b - \theta_a} \tag{7.7}$$

where k_c is the CBOD deoxygenation coefficient (d⁻¹) and L_a and L_b are BOD$_5$ (mg/L) of the water samples collected at two stations with time of travel θ_a and θ_b.

With data shown in Figure 7.7, we have

$$k_c = -\frac{\log 4.6 - \log 3.5}{6\text{-}4} = 0.059 \text{d}^{-1} (\text{base } 10) \tag{7.8}$$

where the calculated CBOD deoxygenation coefficient is in common logarithm or has a base 10. Multiplying a base 10 value by 2.3 transforms it to natural logarithm or has a base e.

$$k_c = 2.3(0.059) = 0.14 \text{d}^{-1} (\text{base } e) \tag{7.9}$$

Comments

(1) L_a and L_b in Eqn. (7.7) are CBOD or ultimate carbonaceous BOD. In practice, BOD_5 is conveniently used as L_a and L_b to measure the field carbonaceous deoxygenation coefficient k_c. Because BOD_5 is closely correlated with CBOD for CBOD ~1.5 BOD_5.

(2) In early days, river water quality modeling was conducted graphically, and base 10 deoxygenation coefficient was used. At present, base e deoxygenation coefficient k_1 is commonly used. Cares should be given when literature values of k_1 are quoted or used to make sure they are all expressed in base e.

Carbonaceous wastes or CBOD in a receiving river may exist in either suspended or dissolved forms. Carbonaceous wastes attached with suspended solids in a river may be removed from river water by settling as well as by deoxygenation. Therefore, immediately downstream from the waste effluent point, the observed rate of CBOD reduction is larger than k_1 and is called as CBOD removal rate k_r, or $k_r = k_1 + k_s$, where k_s is the settling coefficient of suspended solids (Figure 7.8).

Theoretically, in-stream NBOD deoxygenation (nitrification) coefficient can be measured similarly by using in-stream NBOD or ammonia–nitrogen profile (Zison et al., 1978). However, aquatic nitrogen cycle is complex. Sometimes, unidentified NBOD source or sink could exist in a natural river, which could affect the estimation of nitrification coefficient (Brezonik, 1973). Also, nitrifying bacteria may not present in all river reaches. Therefore, precautions are required to make sure the in-stream nitrification indeed takes place when the NBOD profile along a river reach is used to estimate the NBOD deoxygenation coefficient.

7.3.2 Algal bio-productivity

Primary production of algae and other aquatic plants are vital parts of a balanced river ecosystem (Howarth and Michaels, 2000). The photosynthetic and respiration actions of aquatic plants cause the diurnal variation of DO content in a biologically active river. While the rate of respiration

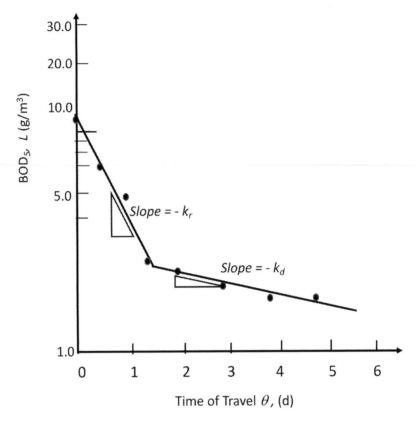

Figure 7.8 Measuring CBOD deoxygenation coefficient and CBOD removal coefficient with data collected by intensive river water quality survey.

R is usually taken as a constant, the photosynthetic DO production rate $P_{DO}(t)$ changes in response to the diurnal variation of solar radiation and other environmental factors. In river DO balance analysis, the rate of DO production by photosynthesis can be expressed as the following sine function (O'Connor and Di Toro, 1970):

$$P_{DO}(t) = (P_{DO})_{max} \sin[\omega(t - t_r)] \qquad t_r < t < t_s$$
$$P_{DO}(t) = 0 \qquad\qquad\qquad\qquad \text{elsewhere} \tag{7.10}$$

where $(P_{DO})_{max}$ is the maximum rate of photosynthetic DO production in a day.

Based on Eqn. (7.10), the following formula relating the average daily photosynthetic DO production rate $(P_{DO})_{ave}$ and the maximum daily photosynthetic DO production rate, $(P_{DO})_{max}$ can be derived as follows:

$$(P_{DO})_{ave} = \frac{\int_0^{T_p} P(t)\,dt}{T_p} = \frac{\int_0^{T_p} (P_{DO})_{max} \sin[\omega(t - t_r)]\,dt}{T_p}$$

$$= (P_{DO})_{max} \frac{2t_f}{\pi} \tag{7.11}$$

The maximum and average daily photosynthetic DO production rates are important water quality parameters. The field experiment can be conducted by using pairs of light and dark bottles attached to a wire cable to estimate the values of these two parameters (Figure 7.9).

During a field survey of stream primary productivity or photosynthetic DO production rate, DO content at light and dark bottle are measured by DO probes (Figure 7.9) at a time period of $(t_1 - t_2)$ during the day-light hours and to yield the following data: initial DO content in the light bottle $(DO_{Light})_{t_1}$; final DO content in the light bottle $(DO_{Light})_{t_2}$; initial DO content in the dark bottle $(DO_{Dark})_{t_1}$; and initial DO content in the dark bottle $(DO_{Dark})_{t_2}$.

Difference of the observed DO increase in the light bottle gives the net DO production rate $(P_{DO})_{net}$ or,

$$(P_{DO})_{net} = \frac{(DO_{Light})_{t_2} - (DO_{Light})_{t_1}}{t_2 - t_1} \tag{7.12}$$

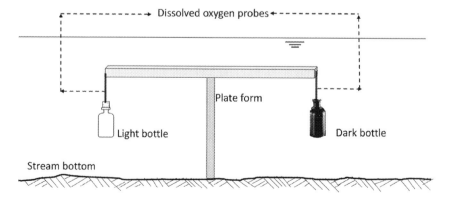

Figure 7.9 Measuring in situ photosynthetic DO production rate by light bottle/dark bottle experiment.

Difference of the observed DO decrease in the dark bottle filled with river water gives the DO consumption rate or $(P_{DO})_{com}$:

$$(P_{DO})_{com} = \frac{(DO_{Dark})_{t_2} - (DO_{Dark})_{t_1}}{t_2 - t_1} \qquad (7.13)$$

$(P_{DO})_{com}$ is the joint result of plant respiration R and DO consumption by organic matters or deoxygenation. In the dark bottle, dissolved organic matters or filtered BOD deoxygenation can be expressed as $(k_1 L_{stream})$, where k_1 is BOD deoxygenation coefficient and L_{stream} is filtered BOD in stream water.

The rate of total photosynthetic DO production over the survey period $(P_{DO})_{photo}$ can then be calculated as follows:

$$\begin{aligned}
(P_{DO})_{photo} &= (P_{DO})_{net} + (P_{DO})_{com} \\
&= \frac{\left[(DO_{Light})_{t_2} - (DO_{Light})_{t_1}\right] + \left[(DO_{Dark})_{t_2} - (DO_{Dark})_{t_1}\right]}{t_2 - t_1}
\end{aligned} \qquad (7.14)$$

The total photosynthetic DO production over the survey period $(P_{DO})_{photo}$ can also be calculated by integration of Eqn. (7.11):

$$(P_{DO})_{photo} = \int_{t_1}^{t_2} (P_{DO})_{max} \sin\left[\left(\frac{\pi}{t_f T_p}\right)(t - t_r)\right] dt \qquad (7.15)$$

By combining Eqn. (7.14) and Eqn. (7.15), the maximum daily photosynthetic DO production rates $(P_{DO})_{max}$ can be determined with survey data:

$$(P_{DO})_{max} = (P_{DO})_{photo} \frac{\pi}{t_f T_p} \left\{ \cos\left[\frac{\pi}{t_f T_p}(t_1 - t_r)\right] - \cos\left[\frac{\pi}{t_f T_p}(t_2 - t_r)\right] \right\} \qquad (7.16)$$

With the $(P_{DO})_{max}$ calculated by Eqn. (7.16), the average daily photosynthetic DO production rates can then be calculated by Eqn. (7.11).

Turbulence of a stream is the primary factor for determining its reaeration capacity. Recent studies also showed that the turbulence in a stream plays an important role in controlling algal bloom (San et al., 2017). Therefore, this study provides useful information for further studies to control the eutrophication and algal boom in a water body by regulating its hydraulic properties.

7.3.3 Reaeration coefficient

BOD deoxygenation coefficient k_1 of a river can be determined in the field by observing the change in the BOD content in a waste plume as it moves downstream. However, the reaeration coefficient k_2 cannot be determined in the same way because the DO content in a stream depends on atmospheric reaeration as well as BOD deoxygenation, algal productivity, and a number of other mechanisms.

7.3.3.1 Measuring reaeration coefficient by tracer techniques

The tracer techniques have been developed to measure reaeration coefficient k_2 of a river. The tracer techniques are used by injecting into the river tracer gases, which are conservative such that their residual concentration in the river water is affected only by water-to-air interface transfer or desorption. The basis of the tracer technique is the observation that the ratio of the desorption coefficient for the tracer gas to escape from water to the absorption coefficient for the oxygen to enter the same water is a constant.

The tracer techniques of the river reaeration coefficient measurement by using Krypton as a conservative gas tracer was first developed by Tsivoglou et al. (1968). After radioactive Krypton gas is injected into a river as a pulse input, normal distributed Krypton gas concentration curves with decreasing peak concentrations are observed in sequential downstream sampling stations. As the decreasing of peak tracer concentrations are also caused by longitudinal dispersion and dilution, another radioactive gas tracer tritium is also injected during the field survey as a dispersion-dilution tracer. The desorption coefficient of krypton-85 can be calculated as follows:

$$k_{kry} = \frac{1}{\theta_{S1-S2}} \ln \frac{\left(\dfrac{C_{kry}}{C_{tri}}\right)_1}{\left(\dfrac{C_{kry}}{C_{tri}}\right)_2} \tag{7.17}$$

where k_{kry} is the desorption coefficient of krypton-85; θ_{S1-S2} is the time of travel from sampling station 1 to sampling station 2; $(C_{kry})_1$ and $(C_{kry})_2$ are the peak krypton-85 concentrations observed at sampling station 1 at sampling station and at sampling station 2; and $(C_{tri})_1$ and $(C_{tri})_2$ are the peak Tritium concentration observed at these two stations.

Generally, the ratio of air–water interface transfer of two different gases under identical stream turbulence is inversely equal to their molecular sizes (Tsivoglou, 1968). The desorption coefficient of krypton-85 as calculated by Eqn. (7.17) can be used to obtain the reaeration coefficient k_2 as follows:

$$k_2 = \frac{k_{kry}}{0.83} \tag{7.18}$$

Krypton is a radioactive gas and can be measured accurately in small concentration, and thus, the amount of krypton injected during a field operation is very tiny and would not produce any public health problem. However, the public objection of using any radioactive material as a tracer would limit its wide application. The US Geological Survey later modified Tsivoglou's method by replacing radioactive tracers with hydrocarbons gases of propane or ethylene (Rathbun, 1977). Tracer methods developed by Tsivoglou and by the US Geological Survey were used in two separate field measurements of reaeration coefficient k_2 in Canandaiqua Outlet, Upstate New York, USA (Tsivoglou, 1974; Rathbun, 1977). Results of these two studies indicate that both methods can yield rather satisfactory results.

Measuring reaeration coefficient by the method of dissolved oxygen balance

Although tracer methods for measuring the river reaeration coefficient yield more reliable results than predictive formulas, they are expensive and difficult to implement (Liu and Fok, 1983). An alternative method of measuring stream aeration coefficient was later derived that calculated the river reaeration coefficient based on field observed diurnal DO variation and bioproductivity (Liu et al., 2015).

Following the Eulerian method of fluid motion, the total rate of change of DO in a one-dimensional river can be described by a substantive derivative of DO deficit D as follows:

$$\frac{dD(x,t)}{dt} = \frac{\partial D(x,t)}{\partial t} + u(x)\frac{\partial D(x,t)}{\partial x} \tag{7.19}$$

Equation (7.19) indicates that the DO content as observed at any particular location in a river is the results of a local change as a function of time and an advective change as a function of flow velocity.

As discussed in Section 7.1, a river water quality modeling is often conducted under critical low flow conditions when the river flow is in a steady state. For a river with a steady-state flow, the advective change of DO is only a function of travel distance. Furthermore, the local change of stream DO is caused mainly by plants photosynthetic oxygen production and respiration (O'Connor and Di Toro, 1970). Therefore, $D(x,t)$ in Eqn. (7.19) can be separated into two components or:

$$D(x,t) = D_\xi(x) + D_\tau(t) \tag{7.20}$$

where $D_\xi(x)$ is a steady-state component of D and $D_\tau(t)$ is a time-varying component of D.

With Eqn. (7.20), the total DO variation in a river can be calculated by the following equation:

$$\frac{dD(x,t)}{dt} = \frac{dD_\tau(t)}{dt} + u(x)\frac{dD_\xi(x)}{dt} \tag{7.21}$$

If the aquatic plants are uniformly distributed along the river, local time-varying of DO or $D_\xi(x)$ can be expressed as follows:

$$\frac{dD_\tau(t)}{dt} = -k_2 D_\tau(t) + [R - P(t)] \tag{7.22}$$

The spatial variation of the DO content in a steady one-dimensional river system is caused by the coupled actions of deoxygenation and reaeration and can be expressed mathematically by the Streeter–Phelps equation.

$$u\frac{dD_\xi(x)}{dx} = k_1 L(x) - k_2 D_\xi(x) \tag{7.23}$$

where $P(t)$ is the rate of plant photosynthesis (g/m³/s).
Substituting Eqns. (7.22) and (7.23) into Eqn. (7.21), we get:

$$\frac{dD(x,t)}{dt} = k_1 L(x) - k_2 D_\xi(x) - k_2 D_\tau(t) + [R - P(t)] \tag{7.24}$$

According to Eqn. (7.20), two reaeration terms in Eqn. (7.24) can be combined as follows:

$$-k_2 D_\xi(x) - k_2 D_\tau(t) = -k_2 \left[-k_2 D_\xi(x) - k_2 D_\tau(t) \right] = -k_2 D(x,t) \tag{7.25}$$

and Eqn. (7.24) becomes,

$$\frac{dD(x,t)}{dt} = k_1 L(x) - k_2 D(x,t) + [R - P(t)] \tag{7.26}$$

In a typical diurnal DO curve, two equilibrium points exist (Figure 7.10). At night-time equilibrium, plant photosynthetic action ceases and the DO deficit is at its maximum or $D = D_{max}$.

$$\frac{dD(x,t)}{dt} = 0 = k_1 L(x) - k_2 D_{max} + R \tag{7.27}$$

At day-time equilibrium, plant photosynthetic action is the most active and the DO deficit is at its minimum or $D = D_{min}$:

$$\frac{dD(x,t)}{dt} = 0 = k_1 L(x) - k_2 D_{min} + \left[R - (P_{DO})_{max} \right] \tag{7.28}$$

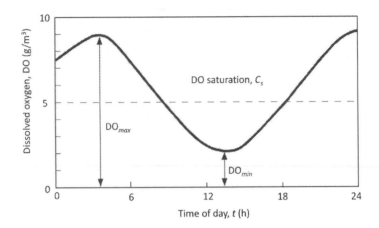

Figure 7.10 A typical diurnal DO curve in a biologically active stream.

where $(P_{DO})_{max}$ is the maximum rate of photosynthetic oxygen production $(g/m^3/d)$.

Combining Eqns. (7.27) and (7.28), a predictive formula of k_2 is derived as follows:

$$k_2 = \frac{(P_{DO})_{max}}{D_{max} - D_{min}} \tag{7.29}$$

Equation (7.28) was used to calculate the reaeration coefficient of Canandaigua Outlet in Upstate New York (see Figure 7.2), where an intensive river water quality survey including direct field measurement of the reaeration coefficient by using tracers was conducted (Tsivoglou, 1974).

The river is relatively flat for the first few miles with an average water depth of 0.5 m, and it becomes much steeper in downstream sections with an average depth of 0.2 m. Using the values of photosynthetic production recommended by USEPA (Zison et al., 1978), the maximum rate of photosynthetic oxygen production $(P_{DO})_{max}$ in Canandaiquae Outlet was estimated to be 17.5 mg/L/d for upper sections and 52.4 mg/L/d for downstream sections. Equation (7.28) was then used to calculate the reaeration coefficient with $(P_{DO})_{max}$ and the maximum and minimum DO points in observed diurnal DO curves. The results are presented in Table 7.2. Also shown in Table 7.2 are the values of reaeration coefficients measured in the field by using the tracer method (Tsivoglou, 1974) and calculated by using empirical formulas derived by O'Connor and Dobbins (1956) and Churchill et al. (1962).

Table 7.2 indicates that the reaeration coefficient in the upper reach from the Outlet to Castle Road as estimated by using the method of DO balance

Table 7.2 Reaeration coefficients (k_2) of Canandaigua Outlet, New York, determined by using the field DO balance method, field tracer techniques, and two predictive equations

River Segments	Field DO Balance Method	Field Tracer Method	O'Connor-Dobbins Formula	Churchill Formula
Outlet to Castel Road bridge	2.43 d^{-1}	2.10 d^{-1}	1.73 d^{-1}	0.54 d^{-1}
Shortsville to Manchester	2.43 d^{-1}	2.43 d^{-1}	3.21 d^{-1}	1.29 d^{-1}

is 2.43 d^{-1} and by using the tracer techniques is 2.10 d^{-1}; and the reaeration coefficient in the lower reach from Shortsville to Manchester as measured by the method of DO balanced is 9.00 d^{-1} and by using the tracer techniques is 9.06 d^{-1}. Therefore, the results obtained by two independent field measurement methods are very close. While both the O'Connor-Dobbins and Churchill equations would significantly underestimate the reaeration capacity in Canandaiqua Outlet.

EXERCISES

1. Results of laboratory analysis of water samples collected at Sampling station 3-Castle Road bridge (Figure 7.2) in a 1978 Canandaiqua Outlet intensive water quality Survey are presented as follows:

Time of Laboratory Analysis (d)	Biochemical Oxygen Demand (g/m^3)
1	3.4
3	8.0
7	17.0
10	32.0
14	40.0
18	46.0
21	47.0
28	49.0

(a) Are the data shown above BOD consumed or BOD remaining? And why?

(b) Use all BOD data points from 1 day to 14 days above to calculate the CBOD deoxygenation coefficient based on the results of laboratory analysis of water sample collected at Sampling Station 3 – Castel Road bridge.

2. Results of the laboratory analysis of water samples collected at Sampling Station 4 – Chapin (Figure 7.2) for Canandaiqua Outlet intensive water quality Survey on 1987 are presented as follows:

Time of Laboratory Analysis (d)	Biochemical Oxygen Demand (g/m³)
1	1.2
3	3.1
5	5.9
7	8.2
9	9.5
14	11.0
19	12.0
21	13.0
28	14.0

Use all BOD data points from 1 day to 14 days from the aforementioned table to calculate the CBOD deoxygenation coefficient based on the results of laboratory analysis of water sample collected at Sampling Station 4 – Chapin.

3. The time of travel between Castle Road bridge station and Chapin station (Figure 7.2) was determined by a tracer study to be 0.75 days.

(a) Use BOD_{28} or CBOD as an index, calculate CBOD deoxygenation coefficient by Eqns. (7.7) and (7.8).

(b) Compare and discuss the CBOD deoxygenation coefficients that were determined in the last two problems by using laboratory long-term CBOD data and that was determined in this problem by using in-stream CBOD distribution.

4. A field survey of the photosynthetic DO production rate using a set of light bottle /dark bottle in a stream was conducted for 8 h (0.33 days) starting at 9:00 am. The following DO variations were observed: initial DO content in the light bottle, $(DO_{Light})_{t1} = 7$ mg/L; final DO content in the light bottle, $(DO_{Light})_{t2} = 12$ mg/L; initial DO content in the dark bottle, $(DO_{Dark})_{t1} = 7$ mg/L; and initial DO content in the dark bottle, $(DO_{Dark})_{2} = 6.5$ mg/L.

Based on the survey results, calculate the average and maximum daily photosynthetic DO production rate.

5. A field survey to determine reaeration coefficient was conducted in a stream by using krypton gas as a tracer and yield the following data:

Sampling Station	Krypton/Ritium Ratio	Time of Travel From the Injection Point, θ (h)
A (Injection point)	0.20	
B	0.15	θ_{AB} = 2.70
C	0.07	θ_{BC} = 2.80
D	0.04	θ_{CD} = 6.10

(a) Calculate and plot the observed concentration ratios of krypton/ tritum along the stream on a semi-logarithm graph paper. Use the log scale for the concentration ratios and the normal scale for the time of travel.

(b) Connect the plotted points with straight lines and estimate the reaeration coefficient of each stream reach between sampling stations.

6. A stream of a 0.58 m deep has a flow velocity of 1.2 m/s. The diurnal DO variation of the stream water was observed and is shown in Figure 7.11:

(a) Determine the stream reaeration coefficient, k_2 by O'Connorand Dobbins formula.

(b) Use Eqn. (7.28) to estimate the maximum rate of photosynthetic oxygen production in the stream, $(P_{DO})_{max}$ (g/m³/d).

(c) Discuss your results.

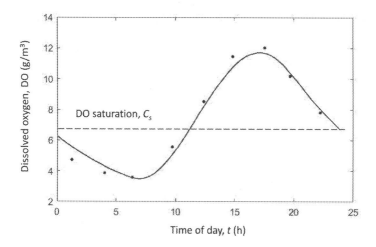

Figure 7.11 Stream diurnal DO variation.

228 Water Environment Modeling

REFERENCES

Brezonik, D.R. (1973). *Nitrogen Sources and Cycling in Waters*. U.S. EPA Research RepEPA-600/3-78-105.

Churchill, M.A., Elmore, H.L. and Buckingham, R.A. (1962). The prediction of stream aeration rates. *Journal of the Sanitary Engineering Division ASCE*, 88(SA4), pp. 1–46.

Howarth, R.W. and Michaels, A.F. (2000). The measurement of primary production in aquatic ecosystems. In *Methods in Ecosystem Science*. Springer, New York, NY, pp. 72–85.

Li, W.H. (1983). *Fluid Mechanics in Water Resources Engineers*. Allyn and Bacon, Inc, Boston.

Linsley, R.K., Franzini, J.B., Freyberg, D.L. and Tchobanoglous, G. (1992). *Water-Resources Engineering*. McGraw-Hill, Inc, New York.

Liu, C.C.K. (1986). Surface water quality analysis. Chapter 1 in Wang, L. and Pereira, N.C. (eds.), *Handbook of Environmental Engineering*, Vol.4. The Humana Press, Clifton, NJ, pp. 1–59.

Liu, C.C.K. and Fok, Y.S. (1983). Stream waste assimilative capacity analysis using reaeration coefficients measured by tracer techniques. *The Journal of the American Water Resources Association* (formerly Water Resour Bull), 19(3), pp. 439–445.

Liu, C.C.K., Lin, P., Xiao, H., Zhang, X., Liu, B. and Zhang, Y. (2015). Evaluation of stream reaeration capacity based on field water quality survey data. *Proceedings of 17th Mainland-Taiwan Environmental Protection Academic Conference*, December 06–10, 2015. Kunming University of Science and Technology Kunming, China.

Mueller, D.S., Wagner, C.R., Rehmel, M.S., Oberg, K.A. and Rainville, F. (2009). *Measuring Discharge with Acoustic Doppler Current Profilers from a Moving Boat*. US Department of the Interior, US Geological Survey, Reston, VA (EUA), p. 72.

O'Connor, D.J. and Di Toro, D.M. (1970). Photosynthesis and oxygen balance in streams. *Journal of the Sanitary Engineering Division*, 96(2), pp. 547–571.

O'Connor, D.J. and Dobbins, W.E. (1956). The mechanism of reaeration in natural stream. *Journal of the Sanitary Engineering Division, ASCE*, 82(6), pp. 1–30.

Rathbun, R.E. (1977). Reaeration coefficients of streams – State of the art. *Journal of the Hydraulics Division, ASCE*, 103(HY4), pp. 409–424.

Rathbun, R.E., Shultz, D.J. and Stephens, D.W. (1975). *Preliminary Experiments with a Modified Tracer Technique for Measuring Stream Reaeration Coefficients*. U.S. Geological Survey, U.S. Geological Survey Open-File Report, pp. 75–256.

San, L., Long, T. and Liu Clark, C.K. (2017). Algal bioproductivity in turbulent water: An experimental study. *Water*, 9(5), p. 304.

Tsivoglou, E.C. (1974). *The Reaeration Capacity of Canandaigua Outlet, Canandaigua to Clifton Springs*. Project No. C5402, New York State Department of Environmental Conservation, Albany, New York.

Tsivoglou, E.C., Cohen, J.B., Shearer, S.D. and Godsil, P.J. (1968). Tracer measurements of stream aeration, II, Field studies. *Journal of the Water Pollution Control Federation*, 40(2), pp. 285–305.

Turnipseed, D.P. and Sauer, V.B. (2010). *Discharge Measurements at Gaging Stations* (No. 3-A8). US Geological Survey, Reston, VA.

Wilson, J.F. (1968). *Fluorometric Procedures for Dye Tracing.* U.S. Geological Survey Techniques of Water-Resources Investigation (No. 3-A12), Reston, VA.

Zison, S.W., Mills, W.B., Deimer, D. and Chen, C.W. (1978). *Rates, Constants, and Kinetics Formulations in Surface Water Quality Modeling.* U.S. Environemntal Protection Agency EPA Research Rep, EPA-600/3-78-105, Athens, GA.

Chapter 8

Modeling of subsurface contaminant transport

8.1 FLOW AND CONTAMINANT TRANSPORT IN SOILS AND GROUNDWATER

Land development in catchment areas requires extensive use of synthetic chemicals such as fertilizers, pesticides, herbicides, and detergents. Residues of these chemicals can cause surface water contamination by rainfall runoff into rivers, lakes, and reservoirs. They can also cause groundwater contamination by polluted seepage water into the soil (Figure 8.1). The flow rate of water in soil and groundwater is much lower than that of river water, and the adsorption of contaminants by soil particles further slows down the moving rate of contaminants. Therefore, it is very difficult to remove contaminants once they enter the soil and groundwater.

8.2 MATHEMATICAL SIMULATION OF FLOW AND CONTAMINANT TRANSPORT IN SOILS AND GROUNDWATER

8.2.1 Transport simulation

The plume of contaminated water in the upper soil moves mainly vertically downward, and it is usually regarded as one-dimensional transport. In addition, natural reactions are often represented by a single first-order decay in model applications (Lin et al., 1995). Therefore, the mass continuity equation of contaminant transport in the soil can be simplified as follows:

$$\frac{\partial C_T}{\partial t} = -\frac{\partial F}{\partial z} - k_b C_T \tag{8.1}$$

where C_T is the contaminant concentration in the soil, F is the mass transport rate of contaminants, and k_b is the first-order decay coefficient of contaminants in the soil.

DOI: 10.1201/9781003008491-8

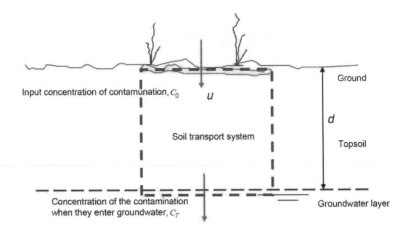

Figure 8.1 Flow and contaminant transport in soils and groundwater.

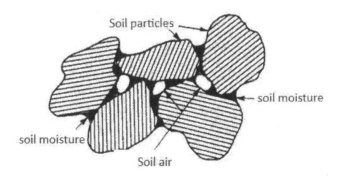

Figure 8.2 The presence of contaminants on topsoil.

Topsoil is usually unsaturated. When contaminated, water plume enters the topsoil with the seepage water, and part of the contaminants in the soil enters the surface of the soil particles or escapes into the soil air due to interface exchange. Therefore, the contaminants can exist in the form of liquid, gas, and solid in the soil (Figure 8.2). The contaminant concentration of the soil (C_T) is the sum of the concentration of liquid, gas, and solid.

$$C_T = e_w C_m + \rho_s C_{TA} + e_a C_g \qquad (8.2)$$

where e_w is the soil porosity occupied by water or soil water content, C_m is the contaminant concentration dissolved in soil moisture, ρ_s is the soil density, C_{TA} is the contaminant concentration adsorbed on the soil particle surface, e_a is the soil porosity occupied by air, and C_g is the gaseous state contaminant concentration in the soil.

When the interface exchange reaches a dynamic equilibrium, the contaminant concentration in the soil air and soil water can be calculated by Henry's law (refer to Section 6.2.1):

$$C_g = K_b C_m \tag{8.3}$$

Similarly, the equilibrium concentration of contaminants on the soil particle surface and in the soil water can be determined by

$$C_{TA} = k_a C_m \tag{8.4}$$

where k_a is the adsorption coefficient. This formula, also known as the adsorption linear isotherm, is commonly used for transport simulation.

Substituting Eqns. (8.3) and (8.4) into Eqn. (8.2), C_T can be represented by C_m as follows:

$$C_T = \left(e_w + \rho_s k_a + e_a K_b\right) C_m \tag{8.5}$$

Usually, the number of contaminants in the gas state is much less than that dissolved in the soil water or adsorbed on the soil particle surface. Ignoring the contaminants in the gas state, formula (8.5) can be rewritten as follows:

$$C_T = \left(e_w + \rho_s k_a\right) C_m \tag{8.6}$$

The transport of contaminants in soil water includes advection and diffusion process. Therefore, the mass transport rate F of the contaminant in Eqn. (8.1) can be written as follows:

$$F = -E\frac{\partial\left(e_w C_m\right)}{\partial z} + u\left(e_w C_m\right) \tag{8.7}$$

where u is the percolating velocity of soil water. The transport of contaminants in the soil is driven by the seepage water; therefore, C_m is the contaminant concentration dissolved in the soil water rather than the total concentration of contaminants in soil C_T.

Substituting Eqns. (8.6) and (8.7) into Eqn. (8.1), we obtain a one-dimensional governing equation for contaminant transport in unsaturated soil layer:

$$\frac{\partial\left(e_w + \rho_s k_a\right) C_m}{\partial t} = \frac{\partial}{\partial z}\left[E\frac{\partial\left(e_w C_m\right)}{\partial z}\right] - \frac{\partial}{\partial z}\left(u e_w C_m\right) - k_b\left(e_w + \rho_s k_a\right) C_m \tag{8.8}$$

The dispersion coefficient and percolating velocity in Eqn. (8.8) are related to the hydrodynamic state in the soil and need to be solved jointly with hydrodynamic equations.

8.2.2 Water flow simulation

The equation for steady flow in porous media is called Darcy's law, for which the mathematical equation is

$$q = -K\nabla H \tag{8.9}$$

where $q = u_x \mathbf{i} + u_y \mathbf{j} + u_z \mathbf{k}$ is the flow velocity vector, K is the hydraulic conductivity, and H is the hydraulic head.

The hydraulic head at any point in water is the sum of its elevation and water pressure, given constant velocity. The water pressure of the water molecules in the unsaturated soil is lower than the atmospheric pressure, and the hydraulic head of the unsaturated soil water can be written as follows:

$$H = z - p_n \tag{8.10}$$

where p_n is the negative pressure, which is called the soil moisture tension or the suction pressure, and its absolute value decreases with the increase of soil moisture content. When the soil moisture is saturated, p_n is equal to zero. The relation curve between moisture tension and water content is an important physical property of soil.

According to the mass continuity equation (2.6), the mass continuity equation for soil moisture content is

$$\frac{\partial e_w}{\partial t} = \nabla \cdot q \tag{8.11}$$

Combining Eqns. (8.9) and (8.11), we obtain

$$\frac{\partial e_w}{\partial t} = \nabla \cdot (K\nabla H) \tag{8.12}$$

Expressing the above equation in the Cartesian coordinate system, we obtain

$$\frac{\partial e_w}{\partial t} = \frac{\partial}{\partial x}\left(K\frac{\partial H}{\partial x}\right) + \frac{\partial}{\partial y}\left(K\frac{\partial H}{\partial y}\right) + \frac{\partial}{\partial z}\left(K\frac{\partial H}{\partial z}\right) \tag{8.13}$$

Equation (8.12) or (8.13) is the hydrodynamic equation for flow in unsaturated soil also known as the Richards equation. When the hydraulic head H in Eqn. (8.13) is expressed by Eqn. (8.10), the one-dimensional Richards equation in the vertical direction can be rewritten as follows:

$$\frac{\partial e_w}{\partial t} = -\frac{\partial}{\partial z}\left(K\frac{\partial P_n}{\partial z}\right) - \frac{\partial K}{\partial z} \tag{8.14}$$

Note here z is positive in the downward direction.

8.3 MODELING OF CONTAMINANT TRANSPORT IN SOILS

8.3.1 Numerical models

The Richards equation is a nonlinear PDE, and its mathematical solution is regarded as one of the major problems in groundwater hydrology. The time and space scales of the Richards equation and the contaminant transport equation are often different in numerical solutions, increasing the difficulty of the simultaneous solution to the two equations. The combination of Richards equations and the contaminant transport equations in two-dimensional and three-dimensional for contaminant transport modeling in unsaturated soil is still a challenging task.

In recent years, studies on the effects of non-uniform distribution of hydraulic properties in soil on contaminant transport have revealed that they can cause macroscopic dispersion of contaminants. The three-dimensional unsaturated soil contaminant transport model based on the Richards equation and the contaminant transport equation has been used to simulate the macroscopic dispersion of contaminants in soil with the log-normal distribution of hydraulic conductivity. The two-dimensional unsaturated soil contaminant transport model has also been used to simulate the transport of pesticide residues in the topsoil, especially the decay reaction of pesticide residues. Both FDM and FEM have been employed in these two-dimensional and three-dimensional models.

The most widely used one-dimensional numerical model for soil contaminant transport is the PRZM model, which is developed by the USEPA (Carsel et al., 1985). It does not apply the Richards equation to simulate soil moisture changes and percolating velocity. Instead, the PRZM model simulates soil hydrodynamic conditions with the use of simple empirical formulas. The contaminant transport equation in the PRZM model is similar to Eqn. (8.8) but includes the reduction of the contaminant concentration in the soil moisture caused by surface runoff, surface soil erosion, and crop absorption.

8.3.2 Analytical models

If the dispersion coefficient, percolating velocity, soil water content, and the first-order decay coefficient are all known constants, the soil transport equation (8.8) can be simplified as follows:

$$\left(1+\frac{\rho_s k_a}{e_w}\right)\frac{\partial C_m}{\partial t}-E\frac{\partial^2 C_m}{\partial^2 z}+u\frac{\partial C_m}{\partial z}=-\left(1+\frac{\rho_s k_a}{e_w}\right)k_b C_m \tag{8.15}$$

The analytical solutions to Eqn. (8.15) can be obtained under special initial and boundary conditions.

By dividing $\left(1+\dfrac{\rho_s k_a}{e_w}\right)$ on both sides of Eqn. (8.15), we obtain

$$\frac{\partial C_m}{\partial t}-\frac{E}{1+\dfrac{\rho_s k_a}{e_w}}\frac{\partial^2 C_m}{\partial^2 z}+\frac{u}{1+\dfrac{\rho_s k_a}{e_w}}\frac{\partial C_m}{\partial z}=-k_b C_m \tag{8.16}$$

Referring to Section 2.4.1, it is found that Eqn. (8.16) has the same form as the most widely used one-dimensional advection–diffusion reaction equation (6.13). The only difference is that the diffusion coefficient and the percolating velocity in Eqn. (8.16) are divided by $\left(1+\dfrac{\rho_s k_a}{e_w}\right)$. As $\left(1+\dfrac{\rho_s k_a}{e_w}\right)$ is a constant greater than 1, this term reduces the percolating velocity and dispersion coefficient in the model equation. Therefore, it is called the retardation factor (RF) in the model analysis.

When the retarded dispersion coefficient and retarded percolating velocity are defined as E_r and U_r, respectively, Eqn. (8.16) becomes

$$\frac{\partial C_m}{\partial t}-E_r\frac{\partial^2 C_m}{\partial^2 z}+U_r\frac{\partial C_m}{\partial z}=-k_b C_m \tag{8.17}$$

After applying pesticides in a farmland, the concentration of pesticides remaining on the soil surface often shows first-order decay and moves downward with the seepage water. Assuming that the downward transport can be simulated with Eqn. (8.17), the initial and boundary conditions can be expressed as follows:

$$C_m(z,0)=0, \ z\geq 0$$
$$C_m(0,t)=C_0 e^{-\gamma t} \tag{8.18}$$

where C_0 is the initial pesticide concentration in source water and γ is the source decay coefficient.

The governing equation (8.17) of the soil contaminant transport model combining the initial and boundary conditions composed of Eqn. (8.18) can be used to obtain the following analytical solution by Laplace transform (see Example 2.1):

$$C_m(x,t) = \frac{C_0}{2}\exp(\alpha x - \gamma t)\left\{\exp\left(-\sqrt{\frac{\beta}{E_r}}x\right)erfc\left(\frac{x}{\sqrt{4E_rt}} - \sqrt{\beta t}\right)\right. $$
$$\left. + \exp\left(\sqrt{\frac{\beta}{E_r}}x\right)erfc\left(\frac{x}{\sqrt{4E_rt}} + \sqrt{\beta t}\right)\right\}$$

(8.19)

where $\alpha = \dfrac{U_r^2}{4E_r} + \dfrac{k_b}{\left(1 + \dfrac{\rho_s k_a}{e_w}\right)}$ and $\beta = \dfrac{U_r^2}{4E_r} + \dfrac{k_b}{\left(1 + \dfrac{\rho_s k_a}{e_w}\right)} - \gamma$.

With various simplified assumptions, the analytical solution model can calculate the long-term contaminant movement in the soil.

Trace pesticide residues were found in groundwater in Hawaii in the early 1980s (Figure 8.3). As groundwater was the main water source for all

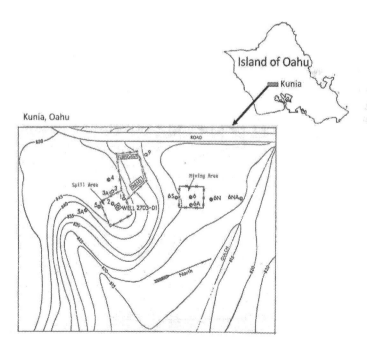

Figure 8.3 The Kunia well with trace residues of pesticides DBCP found and the locations of soil sampling.

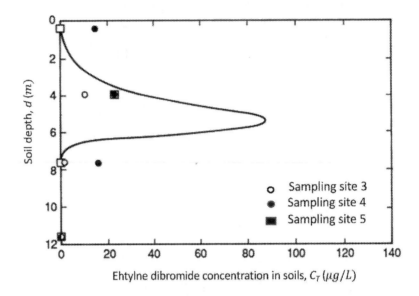

Figure 8.4 Comparisons of calculated and measured residual concentration of dibromoethylene in soil 3 years after the accidental overflow event.

islands in Hawaii, even though the concentration of the contaminants was
very low, it still attracted the attention of the local government and the public. The investigations and studies were quickly carried out to understand
the fate of these pesticides in soil and groundwater (Liu et al., 1983). The
analytical solution model was adopted as an analysis tool.

The coefficients in Eqns. (8.17) and (8.18) were determined by experiments and other studies: percolating velocity u = 0.035 cm/d, diffusion
coefficient E = 0.000024 m²/d, adsorption coefficient k_a = 1.94 mL/g, and
first-order decay coefficient k_b = 1.44 d⁻¹. According to the known coefficients, the dibromoethylene concentration C_m in the soil moisture can be
obtained by solving Eqn. (8.17). Then, the total concentration of dibromoethylene in the soil, C_T, can be calculated by Eqn. (8.6). Figure 8.4 shows the
model calculation results and the measured data at nearby sampling points.

8.3.3   Integrative application of geographic
information system and the simple index
model of contaminant transport in soils

Once the soil and the groundwater are contaminated, it is very difficult
to remove them. Therefore, understanding the potential of groundwater

contamination by pesticides and other harmful chemicals is very important in land development planning. At present, the commonly used one-dimensional transport model and simplified indicator model can only simulate the transport of soil contaminants at a specific location. To apply the simulation results to the planning of regional land development, it is necessary to obtain on-site data of spatial and temporal variation in the region. The GIS is capable of storing information of hydrology, soil, and chemical characteristics of contaminants in an area. Therefore, we can develop an analytical tool for groundwater contamination potential and land development planning by combining GIS and contaminant transport models.

The Index model is a soil contaminant transport model, which is suitable for preliminary analysis of contamination. It considers the soil contaminant transport system (Figure 8.1) as a steady-state plug-flow reactor. The model equation can also be obtained by eliminating the time differential term and the dispersion term in Eqn. (8.17):

$$U_r \frac{\partial C_m}{\partial z} = -k_b C_m \qquad (8.20)$$

Integrating Eqn. (8.20), we have

$$C(z) = C_0 \exp\left(-k_b \frac{z}{U_r}\right) \qquad (8.21)$$

According to Eqn. (8.21), the contaminant concentration decreases with the increase of soil depth, with its rate depending on the decay coefficient and the retarded percolating velocity. If the groundwater layer is below the depth d (Figure 8.1), the contaminant concentration reaching the groundwater layer is

$$C(d) = C_d = C_0 \exp\left(-k_b \frac{d_s}{U_r}\right) \qquad (8.22)$$

The transport of contaminants in the soil can be represented by the RF and the attenuation factor (AF). While the RF is used to calculate the retarded percolating velocity due to adsorption and the time that contaminants enter the groundwater layer from the soil surface, the AF is used to calculate the ratio of the concentration of contaminants entering the groundwater layer

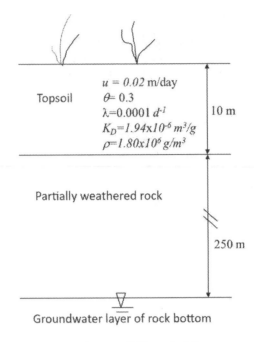

Figure 8.5 Contaminant transport in top soil (Example 8.1).

to the initial concentration. According to Eqn. (8.22), the AF can be defined as follows:

$$AF = \frac{C_d}{C_0} = \exp\left[-k_b\left(\frac{d_s}{U_r}\right)\right] \tag{8.23}$$

Example 8.1 A preliminary study of the contaminant transport in upper soils.

Problem: Deep groundwater is the source of drinking water. There is a topsoil up to 10 m deep on the groundwater layer. The water's downward percolating velocity is 0.02 m per day. The contaminants with a concentration of 50 $\mu g / L$ (parts per billion, ppb) were found on the ground surface. According to the soil and contaminant data shown in Figure 8.5, estimate the seepage time (or residence time) of the contaminant on the topsoil, and the contaminant concentration of the discharged sewage by using the index model.

Solution: (1) Calculate the RF, the retarded percolating velocity, and the residence time of the contaminant in the unsaturated rock formation:

Retardation factor:

$$RF = \left(1 + \frac{\rho_s k_a}{e_w}\right) = \left[1 + \frac{\left(1.8 \times 10^6\right)\left(1.94 \times 10^{-6}\right)}{0.3}\right] = 12.64$$

Retarded percolating velocity:

$$U_R = \frac{u}{RF} = \frac{0.02}{12.64} = 0.0016 \, (\text{m/d})$$

Residence time:

$$\left(t_R\right)_1 = \frac{d_s}{U_r} = \frac{10}{0.00158} = 6250 \tag{d}$$

(2) Calculate the AF and contaminant concentration of the discharged sewage:

Attenuation factor:

$$AF = \exp\left[-k_b\left(\frac{d_s}{U_r}\right)\right] = \exp[-(0.0001)(6250)] = 0.54$$

Contaminant concentration of the discharged sewage:

$$C(d) = C_0(AF) = 50(0.54) = 27 \, (\text{ppb})$$

Comments

The residues of chemical fertilizers and pesticides which enter the soil with seepage water will move toward the groundwater layer, causing groundwater contamination. The index model can calculate the time and contaminant concentration entering the groundwater layer. Therefore, it is an effective tool for assessing the contamination potential. The first-order decay coefficient and the adsorption coefficient have a great influence on the contaminant entering the groundwater layer. If the first-order decay coefficient $k_b = 0.001$ in this example, the contaminant concentration will be reduced from 27 ppb to 0.1 ppb. Therefore, whether the model analysis result is correct depends on whether the model coefficient can be correctly estimated.

Example 8.2 Evaluate soil contamination potential by a conjunctive modeling analysis of GIS and index model.

Problem: The Kaohsiung area, centered on Kaohsiung International Port, Taiwan, is an agricultural area and is also home to large shipyards, steel

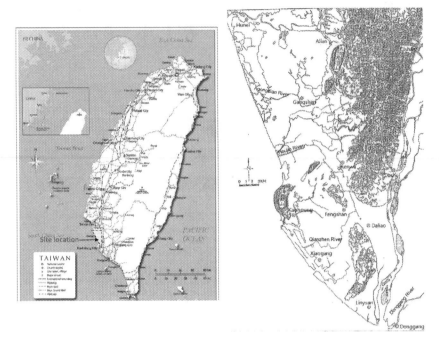

Figure 8.6 Regional location map for groundwater pollution potential assessment in Kaohsiung.

mills, and refinery factories. How to maintain a natural environment and a high-quality living environment for millions of residents in the region during the rapid economic development is a goal for decades. One of the important tasks is to control soil and groundwater contamination. This example is mainly in the Kaohsiung Plain, south of Er-Jen River, and west of Gaoping River (Figure 8.6).

The information used in the GIS in the Kaohsiung local area for contamination potential analysis comes from the groundwater observation wells, i.e., the Water Resources Rainfall Stations of the Taiwan Water Resources Bureau and the Soil Research and Test Center of National Chung Hsing University, Taiwan. Figure 8.7 and Table 8.1 show the distribution and characteristics of 13 soil types in the Kaohsiung region, respectively. Table 8.2 lists the chemical characteristics of contaminants, including commonly used pesticides (herbicides and insecticides) and volatile organic compounds.

The groundwater contamination potential can be assessed by using the AF. The first-order decay coefficient of decomposition of the contaminants

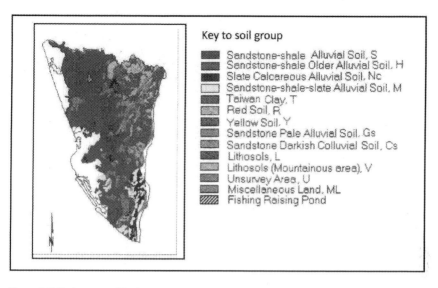

Key to soil group

Sandstone-shale Alluvial Soil, S
Sandstone-shale Older Alluvial Soil, H
Slate Calcareous Alluvial Soil, Nc
Sandstone-shale-slate Alluvial Soil, M
Taiwan Clay, T
Red Soil, R
Yellow Soil, Y
Sandstone Pale Alluvial Soil, Gs
Sandstone Darkish Colluvial Soil, Cs
Lithosols, L
Lithosols (Mountainous area), V
Unsurvey Area, U
Miscellaneous Land, ML
Fishing Raising Pond

Figure 8.7 Soil types in Kaohsiung.

Table 8.1 Soil categories and characteristics in Kaohsiung (south of Er-Jen River and west of Gaoping River)

Characteristics Category	Sand	Silt	Clay	oc	pb (g/cm³)	FC	η	AC
New alluvial soils of sandshale	50.00	37.91	12.09	0.53	1.492	23.82	43.7	19.88
Old alluvial soils of sandshale	50.55	31.08	18.37	0.41	1.476	23.96	44.3	20.34
New alluvial of calcific argillite	0.00	54.74	45.26	2.12	1.206	26.18	54.5	28.32
Mixed alluvial soil of sandshale and argillite	24.40	54.60	21.00	0.68	1.367	24.83	48.4	23.57
Taiwan clay	0.88	53.44	45.68	1.43	1.208	26.15	54.4	28.25
Red clay	14.44	37.35	48.21	1.37	1.253	24.94	52.7	27.76
Loess	37.54	36.07	26.39	0.97	1.405	24.55	47.0	22.45
Pale colluvial soils of sandshale	29.24	55.90	14.86	0.24	1.402	24.52	47.1	22.58
Darkish colluvial soils of sandshale	46.84	38.30	14.86	2.15	1.471	24.03	44.5	20.47
Chisley soil	13.99	57.80	28.21	3.16	1.306	25.38	50.7	25.32
Cliff, bare rock	10.24	58.30	31.46	0.89	1.282	25.52	51.6	25.94
Unsurveyed area	–	–	–	–	–	–	–	–
Reservoir, lake	–	–	–	–	–	–	–	–

Table 8.2 Chemicals considered in contamination potential assessment and their characteristics

Characteristics Category	Koc (cm²/g)	kh	$t_{1/2}$ (d)
Atrazine	372	5.4E-08	49
Aldicarb	36	–	70
Methomyl	11.3	7.5E-09	15
Carbofuran	28	3.1E-07	40
Diuron	342	5.4E-08	217
Saturn	896	6.6E-07	41
Benzene	87	2.4E-01	0.113
Toluene	302	2.8E-01	0.121
Xylene	787	2.2E-01	0.133
Ethyl benzene	832	3.7E-01	0.129
TCE	120	4.2E-01	0.142
Carbon terachloride	347	9.7E-01	0.154

in the soil is represented by k_b, while in this example, it is represented by the half-life $t_{1/2}$ in the GIS, which gives

$$\frac{C_d}{C_0} = \frac{1}{2} = \exp\left(-k_b t_{1/2}\right) \tag{8.24}$$

The value of $t_{1/2}$ can be converted into k_b by using the following formula:

$$k_b = \frac{0.693}{t_{1/2}} \tag{8.25}$$

The rate of downward movement of the delayed contaminant (U_r) is the percolating velocity (u) divided by the RF:

$$U_r = \frac{u}{RF} \tag{8.26}$$

In addition, in the preliminary model analysis, the mean percolating velocity can be estimated by the following formula:

$$u = \frac{q_R}{FC} \tag{8.27}$$

where q_R is the annual average quantity of seepage water and FC is the field water content. If q_R is 500 mm and FC is 23% in a region, the mean percolating velocity is 500/0.23 = 2170 mm/year = 0.006 m/d.

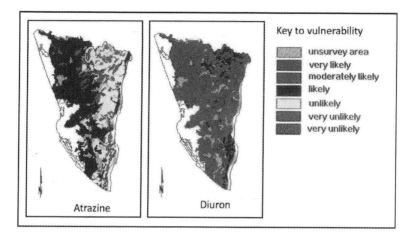

Figure 8.8 Assessment results of groundwater contamination potential in Kaohsiung.

The adjusted retardation equation is

$$AF = \exp\left(-\frac{0.693 \cdot d \cdot RF \cdot FC}{q_R \cdot t_{1/2}}\right) \tag{8.28}$$

In order to assess the groundwater contamination potential in Kaohsiung, the AF is divided into the following ranges:

$0.25 \le AF \le 1.0$, very high contamination risk
$0.1 \le AF \le 0.25$, high contamination risk
$0.01 \le AF \le 0.1$, possible contamination risk
$0.01 \le AF \le 0.01$, low contamination risk
$0.0001 \le AF \le 0.001$, very low contamination risk

Figure 8.8 shows assessment results of the potential for groundwater contamination of pesticide Atrazine and Diuron in the Kaohsiung area. It can be observed that the Diuron has a high contamination potential in groundwater, and the distribution range is dominated by the sandstone shale alluvial soils.

8.4 APPLICATION OF LINEAR SYSTEMS THEORY IN SOIL TRANSPORT MODELING

8.4.1 Response function of a linear systems model

A soil transport linear system can be defined as a contaminated water plume that moves downward along with the seepage flow (Figure 8.9). The variation of water quality in the plume is determined by the amount of

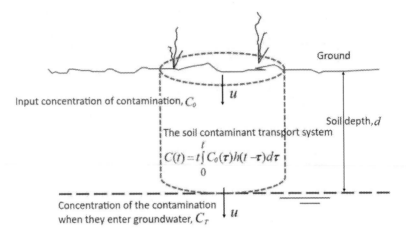

Figure 8.9 Linear system simulation of soil contaminant transport.

contaminant discharge and its self-purification ability. According to the linear system theory (see Section 2.5), the governing equation in the model is the following integral equation:

$$C_{Td}(t) = \int_0^t C_{T0}(\tau)h(t-\tau)d\tau \qquad (8.29)$$

where $C_{T0}(\tau)$ is the contaminant concentration on the topsoil, that is, the input function of the linear transport system; $C_{Td}(t)$ is the contaminant concentration when the seepage water enters the groundwater layer, that is, the output function of the linear transport system; and $h(t-\tau)$ is the impulse response function of the linear transport system.

8.4.2 Compatibility of traditional physically based soil transport model and linear systems soil transport model

The distribution of hydraulic properties in natural soils is usually heterogeneous. The velocity of reaction material moving downward in the soil along with the infiltration flow remains steady at any point, but the infiltration velocity varies with locations. Therefore, as the water and the contaminant dissolved at various locations on topsoil infiltrate downward, the time reaching a fixed depth is different due to different infiltration velocities. This spatial distribution can be expressed by a probability function. Professor Jury at the University of California called this probability function as a transfer function and proposed a transfer function model that simulates the transport of contaminants in heterogeneous soils (Jury, 1982).

Jury further pointed out that the spatial distribution of the downward velocity along infiltration water in the soil can usually be expressed by a log-normal distribution function:

$$h(t) = \frac{\exp\left\{-\dfrac{\left[\ln(t) - \mu\right]^2}{2\sigma^2}\right\}}{\sqrt{2\pi}\sigma t} \tag{8.30}$$

The transfer function is essentially the impulse response function of a linear soil transport system (Liu, 1988). Substituting Eqn. (8.30) into Eqn. (8.29), we obtain a linear system model equation for soil contaminant transport:

$$C_{Td}(t) = \int_0^\infty C_{T0}(\tau) \frac{\exp\left\{-\dfrac{\left[\ln(t-\tau) - \mu\right]^2}{2\sigma^2}\right\}}{\sqrt{2\pi}\sigma(t-\tau)} d\tau \tag{8.31}$$

In general, the analytical solution to Eqn. (8.31) is not easy to obtain and requires numerical integration. However, if $C_{T0}(\tau)$ is a constant C_{T0}, the following analytical solution can be obtained by integration:

$$C_{Td}(t) = C_{T0}\left\{erfc\left[\frac{\ln(t) - \mu}{\sqrt{2}\sigma}\right]\right\} \tag{8.32}$$

If the input function is of rectangular distribution, that is, $C_{T0}(\tau) = C_{T0}$ in the period between 0 and t_1 and $C_{T0}(\tau) = 0$ in the other time, we have

$$C_{Td}(t) = C_{T0}\left\{\int_0^\infty \frac{\exp\left\{-\dfrac{\left[\ln(t-\tau) - \mu\right]^2}{2\sigma^2}\right\}}{\sqrt{2\pi}\sigma(t-\tau)} d\tau - \int_{t_1}^t \frac{\exp\left\{-\dfrac{\left[\ln(t-\tau) - \mu\right]^2}{2\sigma^2}\right\}}{\sqrt{2\pi}\sigma(t-\tau)} d\tau\right\}$$

$$= \frac{C_{T0}}{2}\left\{erf\left[\frac{\ln(t) - \mu}{\sqrt{2}\sigma}\right] - erf\left[\frac{\ln(t-t_1) - \mu}{\sqrt{2}\sigma}\right]\right\} \tag{8.33}$$

Equation (8.33) is an analytical solution of a linear system model for soil contaminants entering with a rectangular input function and then moving downwards. Pesticide or fertilizer is often applied to farmland during a fixed period of time. Therefore, the input function of the soil transport model can be regarded as a rectangular distribution and Eqn. (8.33) can be directly applied in model analysis.

Example 8.3 The Agriculture College of Mexican State University conducted a series of soil contaminant transport experiments in a square field of 8 m on each side (Van de Pol et al., 1977). During the experiment, clean water without contamination infiltrated at a rate of 2 cm/d. After the stable infiltration of the clean water, the conservative chlorides were added as a tracer to the seepage water for the next 36 days. After that, clean water stably infiltrated for 51 days. Three measuring stations were set up in the experiment, and each station was set with eight measuring points at different soil depths. The changes in soil moisture and tracer concentration over time were observed continuously at these points during the experiment. Figure 8.10 shows the change in chloride measured over time at a soil depth of 63.5 cm.

Based on the experimental data, the soil moisture flow rate and the time required to reach any fixed depth can be obtained. Since the water moves at

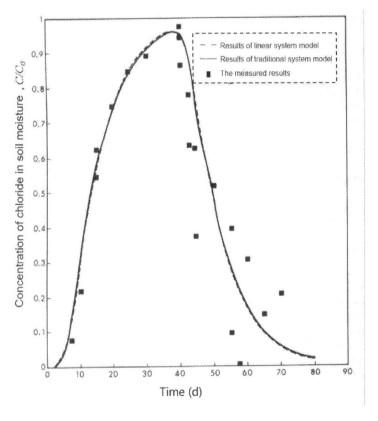

Figure 8.10 Soil transport experiment and simulation results in New Mexico distribution curve of chloride in soil depth 63.5 cm.

a different speed at each point on the soil surface of the experimental field, it takes different lengths of time for water to move downward from the soil surface to a fixed depth. Van de Pol et al. (1977) found that its probability distribution can be expressed by a log-normal distribution function, and the mean value of the log-normal distribution function calculated from the soil surface to the depth of 63.5 cm is 2.631, with the standard deviation of 0.564.

Problem: Simulate the field experiment results of the Agriculture College of Mexican State University using the linear system model and calculate the chloride time curve at soil depths of 63.5 cm and 200.0 cm.

Solution: Since the input is a rectangular function, we can calculate the chloride time curve at a soil depth of 63.5 cm by substituting the parameter value $t_1 = 63.5$ cm, $\mu = 2.631$, $\delta = 0.564$ into Eqn. (8.33). It can be seen from Figure 8.10 that the chloride time curve calculated by the linear system model is close to the measured results.

For the soil with a uniform structure in the vertical direction, the parameter value ($\mu = 2.631$, $\delta = 0.564$) obtained at depth of 63.5 cm can be used at different soil depths. Therefore, when applying Eqn. (8.33) to calculate the chloride time curve of 200.0 cm, we only need to change t_1 to 200.0 cm. The calculation results are shown in Figure 8.11.

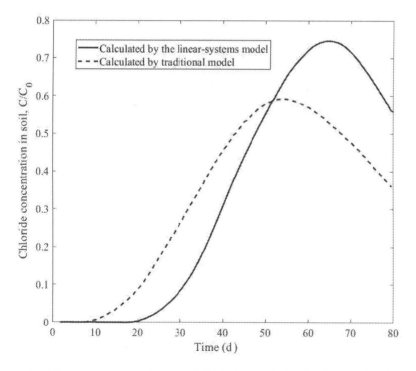

Figure 8.11 Soil transport experiment and simulation results in New Mexico distribution curve of chloride in soil depth 200 cm.

For conservative contaminants, both the adsorption coefficient and decay coefficient are zero. In this case, the one-dimensional soil contaminant transport model equation (8.15) can be written as follows:

$$\frac{\partial C_m}{\partial t} = E \frac{\partial^2 C_m}{\partial^2 z} - u \frac{\partial C_m}{\partial z} \qquad (8.34)$$

According to the experimental design of Example 8.2, the initial and boundary conditions can be expressed as follows:

$$\begin{aligned} &C(0,t) = C_0, \quad 0 < t < t_1 \\ &C(z,0) = 0, \qquad z > 0 \\ &C(\infty,t) = 0, \qquad t > 0 \end{aligned} \qquad (8.35)$$

The analytical solution to Eqns. (8.34) and (8.35) can be obtained by Laplace transform (see Example 2.1).

$$\begin{aligned} C(z,t) = \frac{C_0}{2} &\left\{ \left[erfc\left(\frac{z - ut}{\sqrt{4Et}} \right) + \exp\left(\frac{zu}{E} \right) erfc\left(\frac{z + ut}{\sqrt{4Et}} \right) \right] \right. \\ &\left. - \left[erfc\left(\frac{z - u(t - t_1)}{\sqrt{4Et}} \right) + \exp\left(\frac{zu}{E} \right) erfc\left(\frac{z + u(t - t_1)}{\sqrt{4Et}} \right) \right] \right\} \end{aligned} \qquad (8.36)$$

Equation (8.36) is the analytical solution of a physically based transport model for soil contaminants entering with a rectangular distribution function and then moving downwards.

Example 8.4 Comparisons of physically based and systems soil transport models.

Problem: Using the field measured data in Example 8.2, compare the results obtained by using the analytical solutions to the traditional transport equation model and the linear system model.

Solution: According to continuous observations of soil moisture and tracer concentration obtained at various points in the experimental field, Van de Pol et al. (1977) estimated that the mean percolating velocity is about $u = 3.9$ cm/d, and the diffusion coefficient is about $E = 36.74$ cm²/d. In addition, the input time of the chloride solution in the compound is $t_1 = 36$ days. Substituting these coefficient values into Eqn. (8.36), we can calculate the chloride concentration distribution curve at depths of 63.5 cm and 200.0 cm (Figures 8.10–8.11).

Comments

Both the linear system model derived from the experimental data and the traditional transport model can successfully simulate the transport of contaminants in the soil (Figure 8.10). Applying these two models to predict the transport of contaminants in deeper soils, the calculation results from the linear system model show a relatively large dispersion of contaminants (Figure 8.11). A more detailed discussion on this difference is given in Section 8.4.3.

8.4.3 Conjunctive application of traditional transport model and linear systems model

Both the linear system transport model and the traditional physically based transport model can be used to simulate the contaminants transport in the soil (Figure 8.12). For linear systems, the transport properties of the soil can be represented by an impulse response function, and the results of the transport simulation can be calculated by using the convolution integral of the impulse response function and the input function. For the traditional model, the model equation includes the governing equation, the initial and boundary conditions. The advection–diffusion equation, which is the governing equation of the model, is used to represent the transport properties of a contaminant in soil. The results of the transport simulation can be calculated by solving the model equation.

System identification can be accomplished by using one of the following three methods: the method of physical parameterization, the inverse method, and the method of system parameterization (Liu, 1988). A detailed discussion of these three methods is included in Section 2.5.3. The method of system parametrization to identify the system response function of a linear steady-state river system will be used in the following discussion. The other two methods are applied in watershed system modeling and in the soil transport system modeling as discussed in Chapters 5 and 8, respectively.

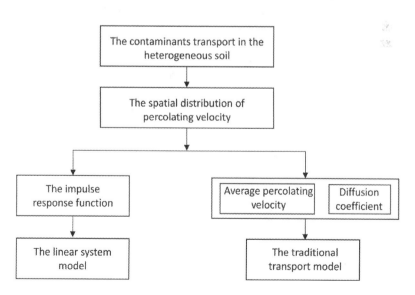

Figure 8.12 Compatibility between linear system model and traditional transport model.

In Section 2.5.1, three methods of the determination of the impulse response function of a linear system or system identification have been introduced, they are as follows: (1) the method of physical parametrization, (2) the method of system parameterization, and (3) the method of system inverse. In Section 5.2, the method of system inverse is used for the system identification of a watershed rainfall–runoff system. In Section 6.5, the method of physical parametrization is used for the system identification of a linear steady-state river system.

The method of system parameterization is used here to determine the impulse response function of a linear soil transport system. When applying this method, it is assumed that some a priori information of system behavior is available such that the impulse response function can be represented by a particular statistical distribution function (Liu, 1988).

As discussed in Section 8.4.1, the spatial distribution of the downward water velocity in a heterogeneous soil can usually be expressed by a log-normal distribution function (Jury, 1982) or the impulse function or transfer function of this soil transport system takes the form of Eqn. (8.30):

$$h(t) = \frac{\exp\left\{-\dfrac{\left[\ln(t) - \mu\right]^2}{2\sigma^2}\right\}}{\sqrt{2\pi}\sigma t}$$

The use of the spatial distribution of the percolating velocity mentioned in Figure 8.12 to estimate the impulse response coefficient is the application of the system parameter method (i.e., gray box method). Equation (8.30) represents the impulse response coefficient expressed by the log-normal distribution function. This distribution function has two parameters, i.e., the mean value μ and the standard deviation σ. Therefore, estimating the impulse response coefficient is equivalent to the estimation of these two parameter values.

Referring to Example 2.2, the impulse response function of a physically based model whose governing equation is Eqn. (8.34) can be written as follows:

$$h(z,t) = \frac{M}{\sqrt{4\pi D_m t}} e^{-\frac{(z-ut)^2}{4D_m t}} \tag{8.37}$$

Therefore, estimating the impulse response function is equivalent to estimating the values of the two physically based parameters, the percolating velocity u and the diffusion coefficient E in Eqn. (8.37).

If the linear system model and the traditional transport model are completely compatible, the impulse response function coefficients represented by

Eqns. (8.37) and (8.30) should be equal. The first moment and the second moment of these two functions with respect to the mean value are obtained first. By letting them equal, the relationships between the parameters in Eqns. (8.37) and (8.30) are as follows (Liu, 1988):

$$u = z \exp\left[-\left(\mu + \frac{\sigma^2}{2}\right)\right] \tag{8.38}$$

$$E = \frac{z^2}{2}\left[\exp\left(\sigma^2\right) - 1\right]\exp\left[-\left(\mu + \frac{\sigma^2}{2}\right)\right] \tag{8.39}$$

By combining Eqns. (8.38) and (8.39), we have

$$E = \frac{zu}{2}\left[\exp\left(\sigma^2\right) - 1\right] \tag{8.40}$$

In Eqn. (8.40), u and σ^2 are constants. Equation (8.40) indicates that the dispersion coefficient of a heterogeneous soil transport system would increase with the increase of soil depth.

Figure 8.11 shows that when applying the system model and the traditional model to predict the transport of contaminant deep into the soil, the range of contaminant distribution curves predicted by the system model is larger than the traditional model. This finding can be explained by Eqn. (8.40). The diffusion coefficient is constant in the traditional model, while Eqn. (8.40) indicates that the diffusion coefficient should increase with the increase of contaminant moving distance.

Gelhar (1986) also found that the dispersion coefficient should increase with the increase of contaminant moving distance through mathematical simulation and statistical analysis of contaminants in heterogeneous soils and groundwater layers and called it as scale-dependent dispersion or macro-dispersion. The study indicates that the dispersion coefficient can be regarded as a constant when the spatial distribution of hydraulic properties in the soil is uniform, but when it is non-uniform or heterogeneous, the macro-dispersion coefficient will increase with the increase of contaminants moving distance.

8.5 MODELING OF CONTAMINANT TRANSPORT IN GROUNDWATER

8.5.1 Numerical models

Groundwater is contaminated by the residual intrusion of various artificial chemicals such as pesticides, fertilizers, heavy metals, and petrochemical materials. Furthermore, seawater intrusion into coastal groundwater is also considered as a contamination problem.

The contaminant transport in groundwater can be modeled by using the advection–diffusion equation (see Section 2.4.1):

$$
\frac{\partial C}{\partial t} = -\frac{\partial (u_x C)}{\partial x} - \frac{\partial (u_y C)}{\partial y} - \frac{\partial (u_z C)}{\partial z}
$$

$$
+ \frac{\partial}{\partial x}\left(E_x \frac{\partial C}{\partial x}\right) + \frac{\partial}{\partial y}\left(E_y \frac{\partial C}{\partial y}\right) + \frac{\partial}{\partial z}\left(E_z \frac{\partial C}{\partial z}\right) + r
\tag{8.41}
$$

Groundwater flow is a flow in porous media. When simulating the contaminant transport of flow in porous media, dispersivity is often used instead of the diffusion coefficient. The diffusion coefficient is the product of the dispersivity and the water flow velocity. While the dispersivity is the property of the porous media and has nothing to do with the nature of the water flow, the diffusion coefficient is related to both.

In the past 30 years, researchers have found that the heterogeneous distribution has a great influence on the contaminant transport. These studies indicate that macro-dispersion strengthens with flow distance when contaminants flow in a heterogeneous porous medium (Gelhar, 1986; Dagan, 1988; Liu et al., 1991; Azimi-Zonooz and Liu, 1994; Neuman and Di Federico, 2003). The macro-dispersion theory is mainly based on the mathematical derivation and statistical analysis, and the biggest problem in applying macro-dispersion to the analysis of groundwater transport model is the lack of measured data required for model verification. In general, it is more difficult to estimate the diffusion coefficient or dispersivity of the groundwater layer than to estimate the permeability and the storage coefficient.

The site where the diffusion coefficient of groundwater layers is estimated is usually composed of an injection well and several observation wells. After the tracer is discharged from the injection well into the groundwater layer, the time-varying concentration of the tracer is measured in the observation well, and then the observed concentration of tracer is compared with the modeling results. The calculation is related to the assumed diffusion coefficient. When the calculation using a certain value of diffusion coefficient is close to the observed value, the diffusion coefficient is adopted. However, because of the slow motion of groundwater flow and the difficulty of capturing flow direction in addition to the presence of macroscopic diffusion, this method is more difficult to apply than that used for determining river diffusion coefficient as introduced in Chapter 6.

8.5.2   Analytical models

The model analysis of groundwater contamination can be achieved by using the simple water quality model. Considering the vertical transport of conservative contaminants in a one-dimensional groundwater layer,

Eqn. (8.41) can be simplified to, by setting the coordinate x pointing downward and $u_x = u$:

$$\frac{\partial C}{\partial t} = -u\frac{\partial C}{\partial x} + E\frac{\partial^2 C}{\partial x^2} \qquad (8.42)$$

The initial and boundary conditions in the groundwater layer are:

$$C = \begin{cases} C_1 & x < 0 \\ 0 & x \geq 0 \end{cases} \quad t = 0$$
$$C = \begin{cases} C_1 & x = -\infty \\ 0 & x = +\infty \end{cases} \quad t > 0 \qquad (8.43)$$

The solution to Eqns. (8.42) and (8.43) can be obtained by the Laplace transformation as follows:

$$C(x,t) = \frac{C_1}{2}\left[1 - erf\left(-\frac{x - ut}{\sqrt{4Et}}\right)\right] \qquad (8.44)$$

This equation indicates that the front of the contaminated plume flows downward along with the groundwater. At the same time, the diffusion will cause the plume to spread in the vertical direction (Figure 8.13).

The most commonly used simple water quality model assumes that the contaminant transport system of groundwater is an ideal PFR. The lateral diffusion coefficient of PFR is infinite, but the longitudinal diffusion coefficient is zero (refer to Section 4.1.3). If the simple model of contaminants in the groundwater layer can be regarded as a steady-state plug-flow reactor, then the contaminants undergo a first-order reaction in groundwater, the governing equation is (refer to Section 4.3.3) as follows:

$$u_e\frac{dC}{dx} = -kC \qquad (8.45)$$

where u_e is the effective velocity of groundwater.

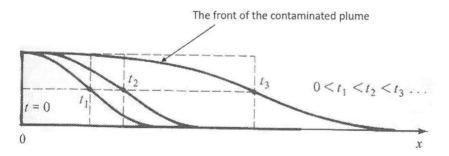

Figure 8.13 Transport of conservative contaminants in one-dimensional groundwater.

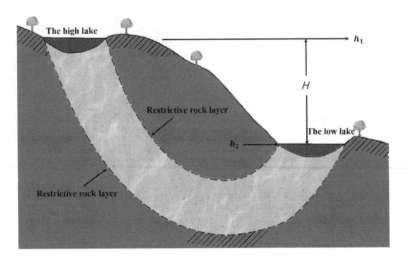

Figure 8.14 Groundwater transport system simulated by plug-flow reactor.

Example 8.5 As shown in Figure 8.14, the bottoms of the two lakes are connected by the closed groundwater layer. The length of the groundwater layer is $L = 2000$ m, the hydraulic head difference is $H = 40$ m, the hydraulic conductivity is $K = 0.001$ m/s, and the effective porosity $e = 0.4$. The high lake is contaminated and the concentration of contaminants is $C_{upper} = 100$ g/m³. This contaminant is first-order decayed in groundwater with a decay coefficient of $k = 0.001$ d⁻¹.

Problem: Assuming the groundwater layer can be considered as an ideal plug-flow reactor, calculate the contaminant concentration C_{lower} of the low lake when groundwater arrives.

Solution: Applying the one-dimensional Darcy's law, the flow velocity of groundwater u is

$$u = -K\frac{dH}{dx} = -(0.001)\left(\frac{-40}{2000}\right) = 0.00002 \text{ m/s}$$

As groundwater can only flow in the porous media, the effective groundwater velocity u_e is

$$u_e = \frac{u}{e} = \frac{0.00002}{0.4} = 0.00005 \text{ m/s}$$

The arrival time of contaminants from high lake to low lake t_{LR} is

$$t_{LR} = \frac{L}{u_e} = \frac{2000}{0.00005} = 4\times10^7 \text{ s} = 463 \text{ d}$$

With the boundary conditions of $x = 0$ and $C = C_{upper}$, the analytical solution to Eqn. (8.45) is

$$C(x) = C_{upper} \exp(-kx/u_e)$$

EXERCISES

1. Figure 8.15 is an ideal PFR for the transport of contaminants in the soil. The reactor length $L = 5$ m (meters). In an experiment, the percolating velocity was $u = 0.12$ m/d, the first-order decay coefficient of contaminants was $k_b = 0.01$ d^{-1}, the contaminant concentration of the inflow water was $C_i = 100$ g/m^3. The mathematical model can be written as

   $$u\frac{dC}{dx} = -k_b C.$$

   (a) Calculate C_{out}, the contaminant concentration of the outflow.
   (b) If the contaminant is absorbed by the porous material with the adsorption coefficient $K_a = 5.8 \times 10^{-6}$ m^3/g, the porous material density $= 2 \times 10^6$ g/m^3, and the porosity $e = 0.25$, calculate the contaminant concentration of the outflow C_{out}.

2. Between the deep groundwater layer and topsoil of Example 8.1, there is a partially weathered rock with a thickness of 250 m (Figure 8.16). The percolating velocity is 0.04 m per day. The adsorption effect for a contaminant in the weathered rock is zero. Other conditions are the same as those in Example 8.1.

   (a) Use the index model to estimate the residence time (infiltration time) of the contaminants in the partially weathered rock and the contaminant concentration of the seepage in the groundwater.
   (b) Discuss the calculation results.

3. When a farmer cleaned the pesticide sprayer, he accidentally poured 10 ft^3 of washing water into a pond. The pesticide residue concentration of the washing water was 100 g/m^3. The pond was 50 ft long, 10 ft wide, and 2 ft deep. The inflow and outflow of the pond were both 5000 ft/d. The pesticide was of first-order decay reaction in the pond with a decay coefficient of 0.01 per day. Because the washing water entered the pond in a very short time, it could be regarded as an instantaneous input (Figure 8.17).

   (a) Assuming the pond as a CSTR, calculate the initial pesticide residue concentration C_0 of the pond after the accident.
   (b) Calculate the change of pesticide residual concentration $C(t)$ in the pond over time. The first-order decay coefficient of pesticide residues in ponds is 0.001/d.

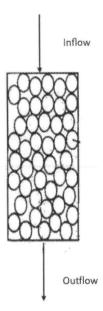

Figure 8.15 Exercise problem 1.

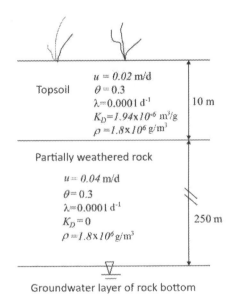

Figure 8.16 Exercise problem 2.

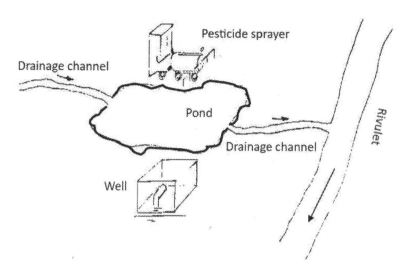

Figure 8.17 Exercise problem 3.

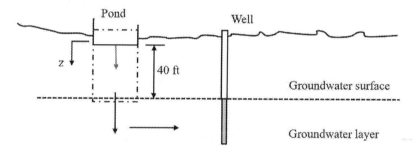

Figure 8.18 Exercise problem 4.

4. Pesticides are not harmful to health but have a strong taste. If the concentration in drinking water is above 0.001 g/m³, it is difficult to swallow. The farmhouse's drinking water is taken from groundwater. The groundwater surface is 40 ft below the ground. The following information was obtained from the preliminary field survey: the percolating velocity from the pond to the groundwater was 15 ft/d, the longitudinal diffusion coefficient was 2.0 ft²/d, and the first-order decay coefficient of pesticide residue in the soil was 0.001/d. At the same time, the pesticide residue concentration in the pond gradually reduced due to the inflow/ outflow of the drainage channel. Because the flow of the groundwater mainly came from the pond seepage and the well was closed to the pond, it could be assumed that the concentration of pesticide residues in the water extracted from the well is the same as the seepage (Figure 8.18).

(a) Review the analytical solutions of the soil contaminant transport models in Section 8.3.2 and write the transport model that can be applied to this example (including the governing equations and initial/boundary conditions, and their analytical solutions).

(b) Calculate how long it will take for the well water to be used again after the farmer finds it too smelly to drink.

REFERENCES

Azimi-Zonooz, A. and Liu, C.C.K. (1994). Stochastic analysis of the relationships between saturated hydraulic conductivity variance and solute dispersion in heterogeneous soils. *Soil Contamination*, 3(3), pp. 225–248.

Carsel, R.F., Mulkey, L.A., Lorber, M.H. and Baskin, L.B. (1985). The pesticide root zone model (PRZM): A procedure for evaluating pesticide leaching threats to groundwater. *Ecology Modeling*, 30, pp. 49–69.

Dagan, G. (1988). Time-dependent macrodispersion for solute transport in anisotropic heterogeneous aquifers. *Water Resources Research*, 24(9), pp. 1491–1500.

Gelhar, L.W. (1986). Stochastic subsurface hydrology from theory to application. *Water Resources Research*, 22(9), pp. 135S–145S.

Jury, W.H. (1982). Simulation of solute transport using a transfer function model. *Water Resources Research*, 18(3), pp. 363–368.

Lin, P., Liu, C.C.K., Green, R. and Schneider, R. (1995). Simulation of 1, 3 dichloropropene in topsoils with pseudo first-order kinetics. *Contaminant Hydrology*, 19, pp. 307–317.

Liu, CC.K. (1988). Solute transport modeling in heterogeneous soils: Conjunctive application of physically based and system approaches. *Contaminant Hydrology*, 3, pp. 97–111.

Liu, C.C.K., Green, R.E., Lee, C.C. and Williams, M.K. (1983). Modeling analysis of pesticide DBCP transport and transformation in soils of Kunia Area in Central Oahu, Hawaii. *Completion Report Submitted to US Environmental Protection Agency, Pacific Biomedical Research Center*. University of Hawaii, 51p.

Liu, C.C.K., Loague, K.M. and Feng, J.S. (1991). Fluid flow and solute transport in unsaturated heterogeneous soils: Numerical experiments. *Contaminant Hydrology*, 7, pp. 261–283.

Neuman, S.P. and Di Federico, V. (2003). Multifaceted nature of hydrogeologic scaling and its interpretation. *Reviews of Geophysics*, 41(3).

Van de Pol, R.M., Wierenga, P.J. and Nielsen, D.R. (1977). Solute movement in a field soil. *Soil Science Society of America Journal*, 41(1), pp. 10–13.

Chapter 9

Estuary, coastal, and marine water modeling

9.1 LINEAR SYSTEMS MODELING OF ESTUARIES

The lower reach of a river under the action of ocean tides is called an estuary or a tidal river. The salinity in a tidal river decreases upstream from that of the ocean water to that of freshwater water (Figure 9.1). In the following presentation and discussion of this section, estuaries and tidal rivers are all referred to as tidal rivers.

9.1.1 Model formulation

The flow velocity and water depth in a tidal river are subjected to spatial and temporal changes continuously. The environmental conditions that produce these changes include upstream river water, tides, wind, earth rotation, etc. The governing equations of tidal-river models are much more complex than non-tidal river models. The derivation and application of comprehensive numerical tidal-river models are discussed by Wang (2002). For a preliminary water quality analysis, a simple analytical tidal river model can be formulated by taking a tidal river as a series of completely stirred tank reactors CSTRs, and each CSTR is simulated as a linear system (Liu et al., 2012). Figure 9.2 shows a tidal-river model consisting of two CSTRs in a series.

The governing equation of a physically based CSTR model takes the form of (refer to Section 4.3.2)

$$\frac{dC_1}{dx} + \lambda C_1 = X(t) \tag{9.1}$$

where subscript 1 denotes the river reach 1, C_1 is the pollutant substance concentration, λ_1 is the characteristic value, $X(t)$ is the input function which can be expressed as $X(t) = L_p(t)/V_1(t)$, and L_p is the pollutant load and V_1 is the volume of the tidal-river reach.

The characteristic value λ_1 in Eqn. (9.1) denotes hydrodynamic and reaction characteristics of a tidal-river reach, which affects its substance mixing and transport. For a simple CSTR system, the hydrodynamic characteristics

DOI: 10.1201/9781003008491-9

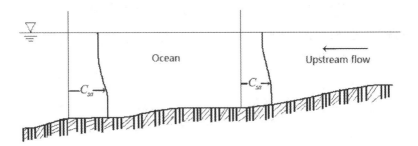

Figure 9.1 Salinity distribution in a well-mixed estuary or tidal river.

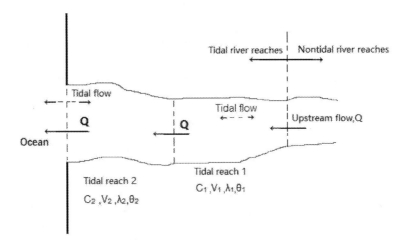

Figure 9.2 Schematic diagram of a simple tidal-river system consisting of two CSTRs.

can be represented by a single-model coefficient, flushing time $\theta_1(d)$. If the relevant reaction of a pollutant substance in the river reach is of first-order decay, the reaction characteristics can be represented by a first-order decay coefficient k_b (1/d). Then, the characteristic value λ_1 in Eqn. (9.1) can be expressed as

$$\lambda = \frac{1}{\theta_f} + k_b \tag{9.2}$$

Note that more terms may be added to the left-hand side of Eqn. (9.2) to represent other relevant hydrodynamic and reaction mechanisms.

The output function of a linear system can be calculated as the summation of a zero-input response and a zero-state response or $C(t) = C_{zero\text{-}input}(t) + C_{zero\text{-}state}(t)$ (Figure 9.3).

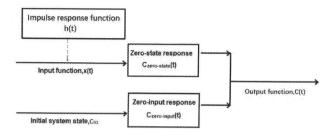

Figure 9.3 Operation of a linear system tidal-river model.

The zero-input response is produced by the initial state of the system. For the CSTR system of the tidal river reach 1 as shown in Figure 9.3, the zero-input response is produced by initial substance concentration C_0. Since all relevant hydrodynamic and reaction mechanisms of the system are represented by the characteristic value of the system λ_1, the zero-state response can be calculated by a simple exponential function,

$$C_{\text{zero-input}}(t) = C_0 \exp(-\lambda t) \tag{9.3}$$

The zero-state response can be calculated by a convolution integration of the system input function $X_1(\tau)$ and the system impulse response function $h(t)$ with $C_{\text{zero-state}}(t) = \int_0^t X(\tau)h(t-\tau)d\tau$. As shown in Section 4.3.2, the impulse response function of a linear system for a physically based system is represented by Eqn. (9.1) is $h(t-\tau) = \dfrac{1}{v}\exp(-\lambda t)$. Therefore, the zero-state response can be expressed as

$$C_{\text{zero-state}}(t) = \int_0^t X_1(\tau)\frac{1}{V_1}e^{-\lambda_1(t-\tau)}d\tau \tag{9.4}$$

With Eqns. (9.3) and (9.4), the linear system model of a CSTR tidal-river system takes the form of

$$C_1(t) = C_{01}\exp(-\lambda_1 t) + \int_0^t X_1(\tau)\frac{1}{V_1}e^{-\lambda_1(t-\tau)}d\tau \tag{9.5}$$

9.1.2 Model application

The linear system model or Eqn. (9.5) is used to simulate a simple tidal-river system consisting of two CSTRs as illustrated in Figure 9.2. When the tidal-river reach 1 receives a constant waste input, the input function becomes $X_1(t) = QC_i$ = constant. Equation (9.5) can be readily solved by integration:

$$C_1(t) = C_{01}e^{-\lambda_1 t} + QC_i \int_0^t e^{-\lambda_1(t-\tau)}d\tau = C_{01}e^{-\lambda_1 t} + \frac{C_i}{\lambda_1\theta_1}\left(1 - e^{-\lambda_1 t}\right) \qquad (9.6)$$

where C_{01} is the initial substance concentration in reach 1.

Note that in a tidal river, the flow in a tidal-river reach as measured by a field survey consists of the upstream freshwater flow Q as well as the oscillating movement of tidal flow (Figure 9.2). As shown in the next section, U.S. EPA (1977) developed a method of the freshwater fraction that can be used to estimate the flushing time.

Similarly, the general form of the linear systems model of the tidal-river reach 2 is,

$$C_2(t) = C_{02}\exp(-\lambda_2 t) + \int_0^t X_2(\tau)\frac{1}{V}e^{-\lambda_2(t-\tau)}d\tau \qquad (9.7)$$

where C_{02} is the initial substance concentration in reach 2 and λ_2 is the characteristic value of reach 2.

As shown in Figure 9.2 that the flow out of reach 1 enters reach 2, or the output function of the reach 1 system $C_1(t)$ is the input function of the reach 2 system or $X_2(t) = Q\,C_1(t)$. The linear systems model of the tidal-river reach 2 becomes

$$C_2(t) = C_{02}\exp(-\lambda_2 t) + Q\int_0^t\left[C_{01}e^{-\lambda_1\tau} + \frac{C_i}{\lambda_1\theta_1}\left(1 - e^{-\lambda_1\tau}\right)\right]\frac{1}{V_2}e^{-\lambda_2(t-\tau)}d\tau \qquad (9.8)$$

Carry out the integration to get

$$C_2(t) = C_{02}\exp(-\lambda_2 t) + \frac{QC_{01}}{V_1(\lambda_2 - \lambda_1)}\left(e^{-\lambda_1 t} - e^{-\lambda_2 t}\right)$$
$$+ \frac{QC_i}{V_2\lambda_1}\left[\frac{\left(1 - e^{-\lambda_2}\right)}{\lambda_2} - \frac{\left(e^{-\lambda_1 t} - e^{-\lambda_2 t}\right)}{\left(\lambda_2 - \lambda_1\right)}\right] \qquad (9.9)$$

A linear system tidal-river model can be derived by taking the system as a series of CSTRs. The model output takes the form of a convolution integration of the system impulse response function and the system input function; the impulse response function characterizes the hydrodynamic and reaction transport mechanisms and the input function defines waste loadings. The model is flexible and easy to apply as the response function and input function can be evaluated independently. Thus, it is a useful analytical tool for preliminary management modeling.

The following example illustrates the model application to estimate the time required by a tidal river to recover after an accidental spill episode.

Example 9.1 Self-purification of a tidal river after an accidental waste spill.

Problem: After an accidental waste spill, a large amount of contaminant substance enters the tidal-river reach 1 and immediately creates an initial substance concentration of 20 g/m³. There is no such pollutant in the entire tidal river prior to the accident. According to a preliminary field survey, this tidal river can be divided into two completely river reaches and simulated by two consecutive CSTR systems, as shown in Figure 9.2. The characteristic values of these two river reaches are $\lambda_1 = 0.1$ d⁻¹ and $\lambda_2 = 0.2$ d⁻¹.

Calculate the change of pollutants concentration in the estuary area.

Solution: Complete mixture in tidal reach 1 and 2 can be assumed.

(1) Tidal-river reach 1:

After the pollution event, the original clean tidal-river reach 1 was immediately polluted with an initial concentration of C_0, and then no pollution entered the tidal reach 1. Therefore, the concentration changes only due to the zero-input response or

$$C_1(t) = \exp(-\lambda_1 t) = \exp(-0.1t)$$

(2) Tidal-river reach 2:

There is no initial concentration in tidal-river reach 2, and the output of tidal reach 1 is the input of tidal reach 2. So, the concentration changes only due to the zero-state response or

$$C_2(t) = \int_0^t C_0 \exp(-\lambda_1 \tau)\exp\left[-\lambda_2(t-\tau)\right]d\tau$$
$$= \frac{\lambda_1 C_0}{(\lambda-\gamma)V}\left[\exp(\lambda_1 t) - \exp(-\lambda_2 t)\right] = \frac{0.1(20)}{0.2}\left[\exp(-0.1t) - \exp(-0.2t)\right]$$
$$= 20\left[\exp(-0.1t) - \exp(-0.2t)\right]$$

The simulation results are shown in Figure 9.4.

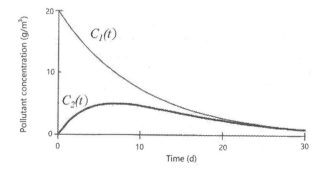

Figure 9.4 Modeling analysis of an accidental spill in a well-mixed tidal river.

After an accidental spill of hazardous wastes in a tidal river system, self-purification of the tidal river has to be evaluated to allow a timely formulation and implementation of a recovery plan. The simple analytical model introduced in this example provides a useful analytical tool for this early-stage evaluation.

9.1.3 Estimating the flushing time of a well-mixed tidal river

The flushing time θ, which characterizes the hydrodynamic behavior of a tidal-river CSTR system, is the time required for freshwater in the system to be replaced. The flushing time of a tidal river is also called the mean hydraulic residence time t_R for a lake CSTR system (see Section 4.3.2). The mean hydraulic residence time t_R of a lake CSTR system with a constant discharge can be readily calculated as V/Q where V is the lake water volume and Q is the discharge. However, the measured flow in a tidal-river reach consists of the upstream freshwater flow as well as the oscillating movement of tidal flow. Therefore, the flushing time of the tidal-river CSTR system with the constant upstream flow is much more difficult to determine than that of a lake CSTR system.

A method of freshwater fraction for estimating the flushing time of a tidal river receiving a constant upstream flow was derived by USEPA (1977). This method can be implemented on the basis of easily obtainable field data and is suitable for preliminary water quality analyses.

The method of a freshwater fraction can be implemented in three steps:
(9) (1) Calculate the fraction of freshwater in a tidal river reach, f_i,

$$f_i = \frac{C_{ss} - (C_{ss})_i}{C_{ss}} \tag{9.10}$$

where C_{ss} is seawater salinity or about 34 $^{0}/_{00}$ and $(C_{ss})_i$ is the average salinity of the ith tidal-river reach.
(2) Calculate the freshwater volume of a tidal-river reach, W_i,

$$W_i = f_i \times V_i \tag{9.11}$$

where V_i is the total water volume of the ith tidal-river reach.
(3) Calculate the flushing water of a tidal-river reach, θ_i,

$$\theta_i = \frac{W_i}{R_i} \tag{9.12}$$

where R_i is the measured freshwater flow of the ith tidal-river reach during a tidal cycle. Note that R_i consists of the upstream flow and the flow from tributaries into the reach (see Figure 9.5).

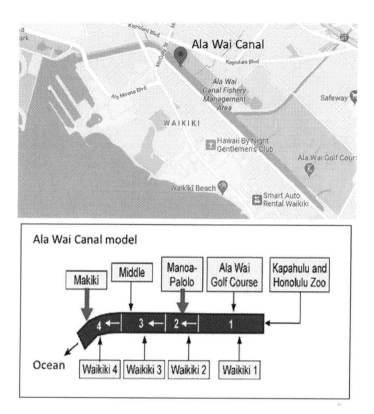

Figure 9.5 A simple water quality model of the Ala Wai Canal on Southern Oahu, Hawaii.

(4) Calculate the flushing water of the entire tidal river by adding the flushing time of individual reaches, $t_f = \sum \theta_i$

Example 9.2 Estimate the flushing time of a tidal river.

Problem: Ala Wai Canal was constructed in 1928 to drain rice paddies and swamps on Southern Oahu, Hawaii, to create dry land for tourism, which would eventually become the Waikiki District. As an artificial waterway, Ala Wai Canal receives the flow from three natural streams and diverts it to the ocean (Figure 9.5). From its upstream end at Honolulu Zoo to the mouth at the Boat Harbor, the Ala Wai Canal is under the tidal influence for its entire length of 3.5 km. During the wet season from October to March, the canal has an average flow of about 49.9 m³/s and the average flow during the dry season from April to September is about 28.1 m³/s.

Calculate the flushing time of the canal during the dry season.

Solution: Based on observed salinity distribution, the canal can be simulated as a series of four completely mixed CSTRs as shown in Figures 9.5 and 9.6. The volume, average salinity, and total freshwater flow rate during

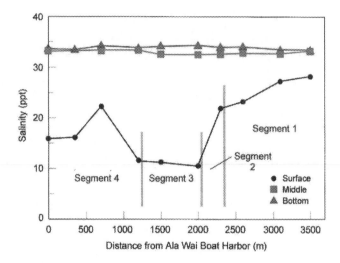

Figure 9.6 Dry season salinity distribution in the Ala Wai Canal.

Table 9.1 The volume, salinity, and the rate of total freshwater flow during a tidal cycle in each of the Ala Wai Canal reaches

CSTRs	Salinity (‰)	Total Volume (m³)	Freshwater Flow (m³/Tidal Cycle)
1	31.49	260,070	13,110
2	24.14	58,740	25,935
3	29.53	172,216	27,567
4	31.74	230,592	35,551

Table 9.2 The flushing time of Ala Wai Canal during dry period as calculated by the method of freshwater fraction

CSTRs	Freshwater Fraction fi	Freshwater Volume W_i (m³)	Flushing Time θ_i (h)
1	0.0885	23,000	21.8
2	0.3012	17,700	8.5
3	0.1452	25,000	11.2
4	0.0812	18,740	6.5

a tidal cycle in each of these reaches are shown in Table 9.1. The flushing time of Ala Wai Canal during the dry period is shown in Table 9.2.

By adding the flushing time of four reaches, the total flushing time of the Ala Wai Canal during a dry season is about 49 h.

Discussion: The method of a freshwater fraction can be used to estimate the flushing time of a tidal river, based on field data of average salinity, freshwater volume, and the rate of total freshwater flow during a tidal cycle. The method provides a reasonable estimation for a preliminary water quality analysis. The estimated total flushing time of the Ala Wai Canal by the method of freshwater fraction as shown in this example is about 49 h. In another independent study that used dye tracer to measure the flushing time, the flushing time of the Ala Wai Canal was estimated to be about 40–60 h (Noda, 1992).

9.2 COASTAL GROUNDWATER MANAGEMENT AND MODELING

9.2.1 Current issues in coastal groundwater management

Many of the world's megacities (>10 million people) are located in coastal areas. Groundwater has been an important source of water supply for municipal and domestic, irrigational, and industrial uses in these areas. A few of these coastal cities have suffered from groundwater mismanagement. The most common issues in coastal groundwater management include land subsidence and seawater intrusion induced by groundwater over exploitation, as well as the determination of sustainable yield (sometimes termed as safe yield) of coastal aquifers. The management of groundwater resources in coastal areas is a difficult task. The objective of this chapter is to provide basic information about coastal groundwater systems and their management.

9.2.1.1 Land subsidence

Coastal zones that are prone to land subsidence are those with extensive subsurface layers of clay and peat. Roughly, a half billion people live in delta regions threatened by land subsidence (Syvitski et al., 2009). The leading cause is the loss of structural support of the rock grains in the subsurface, which is primarily due to the removal of groundwater by excessive pumping (USGS, 2016). A lowering of the land surface relative to sea level can be a trigger for seawater intrusion.

Rapidly expanding urban areas require huge amounts of water for domestic and industrial water supply, which often leads to over-exploitation of groundwater resources. In the city Dhaka in Bangladesh for instance, continuous large-scale extraction currently causes groundwater levels to fall by 2–3 m/y (Deltares, 2015). The extraction of groundwater causes severe land subsidence, and similar conditions occur in other coastal cities like Tianjin (China), Jakarta (Indonesia), Ho Chi Minh City (Vietnam), and Bangkok (Thailand). The current global mean sea level rise is around 3 mm/y. This rate is expected to increase in the future (IPCC, 2013), but still it remains rather small when compared with subsidence rates of 6–100 mm/y in some coastal megacities.

9.2.1.2 Seawater intrusion

Seawater intrusion is the displacement of freshwater in a coastal aquifer by seawater. The cause of seawater intrusion can be natural, for example, a decrease of recharge or a rise in local sea level, but in most cases, aquifer over-exploitation has been the key factor.

Aquifers in inland-area exist in the form of confined and unconfined aquifers. For the unconfined aquifers, the surface is the groundwater surface, and the bottom is the confining rock bed with very low permeability; for confined aquifers, the top and bottom are both the confining rock beds. In contrast, most of the aquifers in the coastal area exist in the form of basal aquifers. The freshwater in the basal aquifer floats on the sea water, so there is no fixed bottom for basal aquifers (Figure 9.7).

When the groundwater flow in a basal aquifer is in a steady condition, the transition zone between the freshwater and seawater layers is relatively narrow and can be taken as a sharp interface (Figure 9.7). In order to keep the interface in a fixed position, the pressure on both sides of the interface must be equal or

$$\rho_f (H + \zeta) = \rho_s \zeta$$
$$\zeta = \frac{\gamma_f}{\gamma_s - \gamma_f} H \approx 40H \tag{9.13}$$

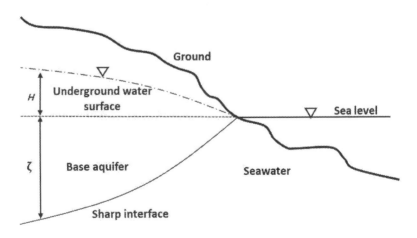

Figure 9.7 Schematic diagram of basal aquifer.

where H is the freshwater head, ζ is the basal aquifer thickness, the specific gravity of freshwater γ_f is about 1.00, and the specific gravity of sea water γ_s is about 1.025.

Equation (9.13) indicates that if the thickness of the freshwater layer in the basal aquifer is h above the sea level, the thickness of the freshwater layer below the sea level is 40 H (Figure 9.7). This equation is also known as Ghyben–Herzberg equation, in memory of the two hydrologists who first deduced this equation.

In coastal areas, over-exploitation of groundwater will lead to the decline of the freshwater head. According to the Ghyben–Herzberg equation, if the freshwater head drops 1 m, the sharp interface would rise 40 m. Consequently, the original freshwater layer is replaced by seawater, causing the problem of seawater intrusion.

In the above discussion, it is assumed that the bottom of the basal aquifer is a sharp interface, that is, the freshwater becomes the seawater in a very short distance. In fact, there is a transition zone between the freshwater and seawater layers, where the salinity gradually decreases from the bottom to the top (Figure 1.7). Before exploitation, the transition zone of the basal aquifer is usually very thin, and its bottom can be assumed to be of a sharp interface. The research found that over-exploitation of basal groundwater will not only cause the decline of the freshwater head but also expand the transition zone (Liu et al., 1991). The World Health Organization stipulates that the salinity of drinking water shall not exceed 4% of the salinity of seawater. By this standard, most water in the transition zone is unusable. Therefore, seawater intrusion in coastal areas must be considered from two aspects: the decline of groundwater level and expansion of transition zone.

9.2.1.3 Sustainable yield

All groundwater pumping comes from capture. The greater the intensity of pumping, the greater the capture. Capture comes from decreases in natural discharge and increases in recharge. Natural discharge supports riparian, wetland, and other groundwater-dependent ecosystems, as well as the baseflow of streams and rivers. Capture depends on usage, and it is not related to size or hydrogeological characteristics of the aquifer, or the natural recharge. Excessive groundwater pumping can lead to groundwater depletion, and this may have serious social and economic consequences.

Attempts to limit groundwater pumping have been commonly based on the concept of safe yield, defined as the attainment and maintenance of a long-term balance between the annual amount of groundwater withdrawn by pumping and the annual amount of recharge. This definition is regarded as 'narrow' because it does not consider the needs of groundwater-fed surface water (springs and baseflow) and groundwater-dependent ecosystems (wetlands and riparian vegetation) (Sophocleous, 1997). More recently, the emphasis has shifted to sustainable yield (Alley and Leake, 2004; Maimone, 2004; Seward et al., 2006).

Sustainable yield is the volumetric rate of water that an aquifer can provide while maintaining desirable aquifer conditions (primarily water levels and water quality) such that the aquifer remains a viable resource for long-term use. An aquifer managed under sustainable-yield conditions is a self-renewing resource. Sustainable yield is a critical element in identifying and designing viable water supply alternatives. In a natural state, the groundwater in the coastal area flows slowly into the ocean, while the rainfall infiltrates into the aquifer continuously. There is a good balance between outflow and inflow, so the capacity and water quality of the underground reservoir can be kept stable (Figure 9.7). A large number of groundwater extraction changes this nice natural state of balance and will cause the formation of subsidence and seawater intrusion and other adverse consequences. The gravity load borne by the closed aquifer from the upper part is shared by the porous material and water in the aquifer (Figure 9.7). Overextraction of groundwater will lead to the decrease in water head and the increase in pressure exerted on the porous media and consequently resulting in land subsidence. The basal aquifers of Hawaii are composed of porous basalt, so there is no threat of land subsidence. Under such conditions, seawater intrusion is the most important factor that influences the sustainable yield.

9.2.2 Sharp interface approaches for seawater intrusion

In most cases, coastal groundwater modeling needs to deal with variable density fluid or two fluids of different densities separated by an interface (e.g., freshwater and sea water). Depending on the kind of situation that exists, two different approaches are often used, namely, the sharp interface method and variable density models. Solutions with a sharp interface are easier to obtain but there are limitations on their physical representation.

Though freshwater and saltwater tend to mix, under steady flow conditions, regular distribution of salinity with a steady mixing zone between freshwater and saltwater is possible. In many real situations, the thickness of this mixing zone is small relative to that of the fresh and sea water bodies. This is the well-known basis for the sharp interface approximation of flow in coastal aquifers, which dramatically simplifies the problem by considering two immiscible fluids of different densities separated by an interface presenting similar properties as a water table boundary. This approximation, together with the equilibrium of pressures on the interface (generally using the hydrostatic formula of Ghyben–Herzberg) and the Dupuit–Forchheimer assumptions of homogenous horizontal flow, lead to well-known approximate analytical solutions. They have been widely used with success in simple situations or preliminary assessments. Many of these analytical solutions can be found in Bear (1979), Huisman (1972), and Custodio et al. (1987).

The two-fluid situation in a coastal aquifer or an island where freshwater and saltwater can be assumed separated by a sharp interface can easily be formulated as two differential equations coupled by the interface movement condition and the pressure equilibrium on it (Hubbert condition). When there is a very thick saltwater body, saltwater flow can be neglected and the problem is reduced to solving only the freshwater part under appropriate boundary conditions. Boundary conditions are relatively easy in islands below which there is a continuous saltwater body. This is a classical problem (e.g., Guswa and LeBlanc, 1980; Ayers and Vacher, 1983; Contractor, 1983).

Challenges occur when a transient two-dimensional solution (in horizontal coordinates) is sought for the freshwater body alone, and saltwater movement is significant. The real interface cannot move as fast as the Ghyben–Herzberg condition assumes. For freshwater reduction due to abstraction, improvements of prediction have been sought by using the freshwater depth sometime earlier, thus introducing an artificial time lag. However, there is no theoretical basis for estimating this time lag, even though it seems to yield good results in particular cases. The consideration of the saltwater body notably improves the solution, but this is more complicated and time-consuming.

We now use an example to illustrate how the sharp interface approach can be used in coastal groundwater analysis. The groundwater level of Pearl Harbor was decreasing yearly due to large-scale pumping (Figure 9.8). In order to protect water sources, surveys and research have been conducted

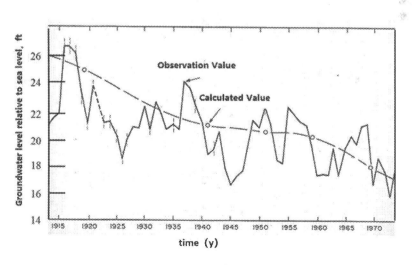

Figure 9.8 Time history of groundwater level of the Waipahu observation well (Well no. 2300.10) in the Pearl Harbor.

to develop and implement various management schemes. At an early stage, coastal groundwater modeling often assumed there was a sharp interface between freshwater and sea water (Liu et al., 1983; Essaid, 1990).

The governing equation of the two-dimensional basal aquifer model can be written as

$$\frac{\partial}{\partial x}\left(d_g K_x \frac{\partial H}{\partial x}\right) + \frac{\partial}{\partial y}\left(d_g K_y \frac{\partial H}{\partial y}\right) = f_r \frac{\partial H}{\partial t} + S_t \tag{9.14}$$

where d_g is the thickness of the groundwater layer. According to the Geyben–Herzberg equation, $d_g = \zeta + H$; K_x and K_y are hydraulic conductivity coefficients; and f_r is the retention coefficient and S_t is the source or sink.

Figure 9.9 shows the grid system and boundary conditions of the Pearl Harbor basal aquifer. About 100 pumping wells were divided into 18 pumping areas. In groundwater modeling, the establishment of initial conditions is regarded as a part of model calibration. In this case, the groundwater model was used to reconstruct the basal aquifer of Pearl Harbor. First, the

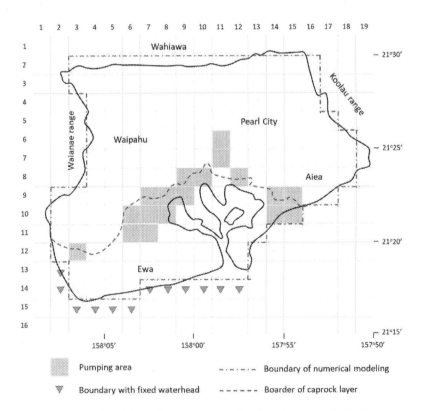

Figure 9.9 Numerical modeling of groundwater in Pearl Harbor using the FDM.

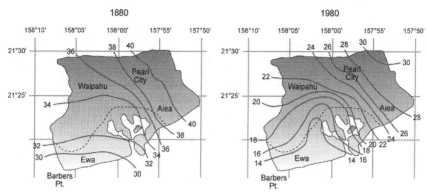

Figure 9.10 Spatial distribution of groundwater head in Pearl Harbor calculated by the numerical model.

groundwater level was set to zero, and the natural infiltration induced by rainfall was trigged. The water level of the groundwater and the discharge at the coastline boundary (caprock layer) increased gradually. When the groundwater head and discharge reached a stable state, the numerical reconstruction of Pearl Harbor's groundwater was completed. Groundwater pumping at Pearl Harbor started in 1880. The hydraulic head contour of 1880 was given in Figure 9.10(a), which was also the initial condition of the model. The amount of water pumped since 1880 was well-documented. The actual groundwater extraction data were adopted into the numerical model, and the variation of water level (the thickness of basal aquifer) can be calculated. Figure 9.10(b) shows the calculated results of groundwater head distribution in 1980.

9.2.3 Variable density models for seawater intrusion

Coastal aquifers considering the existence of a finite length of mixing zone can be simulated by using variable density models. For such cases, there exist rare analytical solutions except for very simple situations (e.g., Henry, 1964), which have been widely used as a test problem (Segol et al., 1975; Frind, 1982). Volker et al. (1985) use an approximate analytical solution to study an island.

The density-dependent solute transport approach requires the simultaneous solution of two PDEs (flow and transport). Density is related to the concentration of salt by certain empirical equations. Generally, vertical 2D approaches have been applied to study the difference between the variable density and the sharp interface solutions, such as the works of Segol and

Pinder (1976) and Volker and Rushton (1982). The latter also considers the sensitivity of the solution to the change of dispersivity.

Different numerical codes are available for variable density situations. Models considering aquifer heterogeneities, such as aquitards, may be very complicated for practical uses. Thus, simplifying assumptions are often needed. The effect of aquitards under some conditions can be treated analytically (Frind, 1982), thus improving the model performance in practice.

Variable density approaches still have their limitations. When the mixing zone is thin, the concentration gradient across it is very high and this may lead to numerical instabilities unless the mesh is very fine in this zone. In such situations, the sharp interface approach yields similar results more easily and without unknown numerical problems. The toe and tip problems do not appear when using a variable density model, but numerical dispersion may unrealistically enlarge the mixing zone near them.

SUTRA code is a mathematical model developed by the US Federal Geological Survey to analyze seawater intrusion into groundwater (Voss, 1984ª). SUTRA adopts the FEM to discretize the governing equations of groundwater, including the mass balance equation and solute transport equation. The temporal and spatial variation of water head, solute concentration, and water temperature in the saturated and unsaturated regions can be modeled.

In the solute transport equation, the flow velocity and diffusion coefficient change with the hydrodynamic properties. Therefore, the hydrodynamic simulation is usually carried out first, and then the resultant flow velocity and diffusion coefficient are inputted into the transport model. In the case of seawater intrusion, the salinity changes greatly, which leads to the change of water density, and ultimately the hydrodynamic properties. Therefore, for seawater intrusion problems, the hydrodynamic model and solute transport model must be solved simultaneously. The hydrodynamic equation of the SUTRA model is (Voss, 1984a)

$$\left(S_w \rho S_{op} + e\rho \frac{\partial S_w}{\partial p}\right)\frac{\partial p}{\partial t} + \left(eS_w \frac{\partial \rho}{\partial C_{sa}}\right)\frac{\partial C_{sa}}{\partial t} - \nabla \cdot \left[\left(\frac{k_p \rho}{\mu}\right)(\nabla p - \rho g)\right] = Q_p \quad (9.15)$$

where S_w is the water saturation, S_{op} is the specific pressure storativity, C_{sa} is the salinity, e is the porosity, k_p is the permeability of the porous medium (only related to the geometric properties of the porous medium), and Q_p is the groundwater recharge.

The solute transport equation for SUTRA mode is

$$eS_w \rho \frac{\partial C}{\partial t} + eS_w \rho q \nabla C - \nabla \cdot (eS_w \rho E \nabla C) = Q_p \left(C_{sa}^* - C\right) \quad (9.16)$$

where q is the flow velocity in Darcy's law, E is the dispersion coefficient, and C_{sa}^* is the salinity of the recharged water.

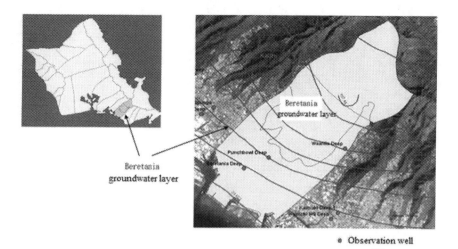

Figure 9.11 Beretania groundwater layer and observation station location on Oahu Island, Hawaii.

Equations (9.15) and (9.16) are coupled in the following fashion: (1) the water pressure calculated by Eqn. (9.15) is needed when calculating the flow rate in Eqn. (9.16) using Darcy's law; (2) the salinity calculated by Eqn. (9.16) is needed when calculating the distribution of water pressure by using Eqn. (9.15).

Liu et al. (1991) applied the vertical two-dimensional SUTRA model to simulate the basal aquifer of Beretania, south of Oahu Island, Hawaii (Figure 9.11). The results showed that the transition zone between freshwater and seawater expanded in response to groundwater exploitation. The study estimated hydraulic conductivity, solute transport, and other model parameters. The field survey was performed, and the measured data were used for model calibration. The numerical results showed that increasing the pumping amount from the basal aquifer would cause the decrease of groundwater head, as well as the expansion of the transition zone. Therefore, a detailed study on the freshwater and seawater mixing mechanism in the transition zone at the bottom of a basal aquifer is very important for groundwater resources management.

Gingerich and Voss (2005) used a three-dimensional changing-density groundwater model to simulate the coastal aquifer in the south part of Oahu, Hawaii, USA (Figures 9.12 and 9.13). Based on the computational results of groundwater flow and solute transport, it was found that the adjustment of the transition zone at the bottom of the basal aquifer takes quite a long time. After the basal aquifer went through pumping, irrigation, recharge, etc., it would take about 50 years for the transition zone to move to its

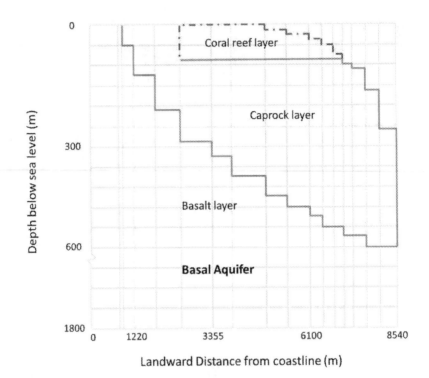

Figure 9.12 Finite element modeling of Beretania basal aquifer using SUTRA.

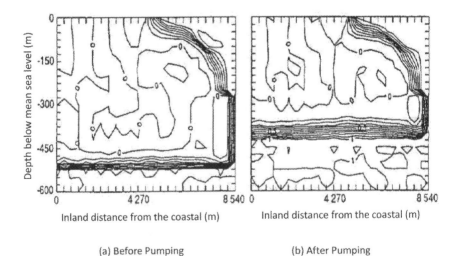

Figure 9.13 Numerical modeling of salinity transport and enlargement of transition zone using SUTRA.

new stable position. The numerical results indicated that the transition zone moves upwards and toward the land. The numerical model employed was 3D SUTRA, and some new tools were adopted for pre- and post-processing and three-dimensional dynamic visualization.

Figure 9.13 Numerical modeling of salinity transport and enlargement of transition zone using SUTRA: (a) before pumpimg and (b) after pumpOne of the difficulties in modeling the three-dimensional basal aquifer is the estimation of model coefficients. Two publications conducted the three-dimensional modeling of the basal aquifer of Oahu Island, Hawaii, in which Gingerich and Voss (2005) estimated that the effective porosity and lateral diffusivity of Pearl Harbor are 0.04 and 0.82 ft, respectively, while Todd Engineers (2005) found these coefficients to be 0.20 and 16.4 ft, respectively. Although the coefficient values used were quite different, both publications considered that the model calibration has been completed (Liu, 2007). Therefore, before using these complex models in groundwater management, the estimation of model coefficients should be more consistent.

When pollutants flow in the heterogeneous porous medium, the macroscopic diffusion will increase with the flow distance. Due to macroscopic diffusion, it is more difficult to estimate the diffusion coefficient or diffusion degree of groundwater layers than to estimate the hydraulic conductivity and storage coefficient. A method for measuring the diffusion coefficient of the basal aquifer of Hawaii will be introduced in Section 9.2.4 (Liu and Dai, 2012).

9.2.4   Analytical models for sustainable yield

Sustainable yield is one of the key indexes in coastal groundwater management. In this section, we use RAM2, an analytical model, as an example to illustrate the evaluation of sustainable yield in coastal areas. RAM2 including two sub-models: hydrodynamic model and salinity transport model (Figure 9.14).

9.2.4.1   Derivation of analytical coastal groundwater management model RAM2

If the transition zone of a basal aquifer can be viewed as a sharp interface, the flow in the aquifer can be simplified as a one-dimensional flow (Figure 9.15). According to Gehben–Herzbery's law, the depth of the sharp interface is about 40 times of the hydraulic head (Lau and Mink, 2006). Therefore, the thickness of the basal aquifer is $\sigma = 40H + H = 41H$, as in Figure 9.15.

The governing equation for the one-dimensional groundwater flow can be written as follows:

$$\frac{\partial (f_s H)}{\partial t} = \frac{\partial (-q)}{\partial x} + W_Q (x,t) \qquad (9.17)$$

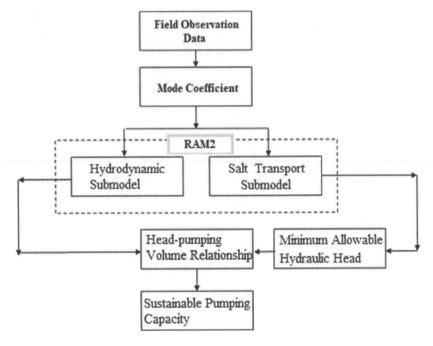

Figure 9.14 RAM2 model and sustainable yield evaluation.

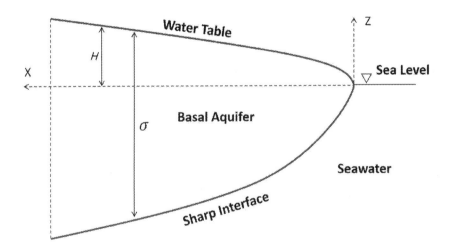

Figure 9.15 Flow system diagram of RAM2 model.

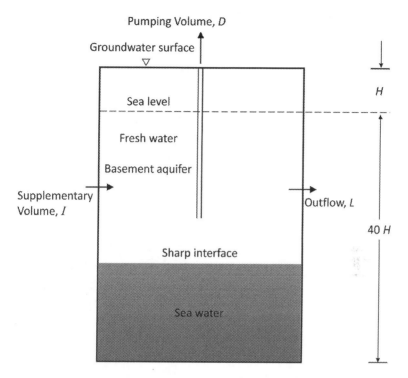

Figure 9.16 The basal aquifer simulated by a complete hybrid ideal reactor.

where f_s is the storage coefficient, W_Q is the amount of inflow and outflow per unit area, x is the flow distance (m), and $q = -41K\dfrac{\partial H}{\partial x}$ by Darcy's law.

Mink (1981) pointed out that the outflow can be written as $L = (H/H_0)^2 I$, where H_0 is the initial head and I is the amount of inflow or natural infiltration. Assuming that the unit-width discharge q does not change with x, Eqn. (9.17) can be simplified as

$$41SA\frac{dH}{dt} = \left[1 - \left(\frac{H}{H_0}\right)\right]I - D \tag{9.18}$$

Mink (1981) derived Eqn. (9.18) from the one-dimensional hydrodynamic equation. If the basal aquifer is regarded as a well-mixed ideal reactor (Figure 9.16), Eqn. (9.18) can also be directly derived.

When the infiltration amount is constant, the analytical solution of Eqn. (9.18) is (Liu and Dai, 2012)

$$d_{i-1}=d_0\sqrt{\frac{I-D}{I}}\frac{\left[\sqrt{I-D}+\dfrac{d_i}{d_0}\sqrt{I}\right]\exp\left[\dfrac{2\sqrt{(I-D)I}\left(t_{i-1}-t_i\right)}{V_0}\right]-\sqrt{I-D}+\dfrac{d_i}{d_0}\sqrt{I}}{\left[\sqrt{I-D}+\dfrac{d_i}{d_0}\sqrt{I}\right]\exp\left[\dfrac{2\sqrt{(I-D)I}\left(t_{i-1}-t_i\right)}{V_0}\right]+\sqrt{I-D}-\dfrac{d_i}{d_0}\sqrt{I}}$$
(9.19)

Let the time approaches infinite, the steady-state solution of Eqn. (9.18) can be obtained:

$$\frac{H}{H_0}=(1-n)^{1/2}$$
(9.20)

where $n = D/I$ is the ratio of extraction amount to natural infiltration amount.

Mink (1980) found that the dimensionless sustainable yield (D/I) is related to the dimensionless head (h_e/H_0), as shown in Figure 9.17. The minimum equilibrium head h_e can be estimated by an empirical formula (Mink, 1980).

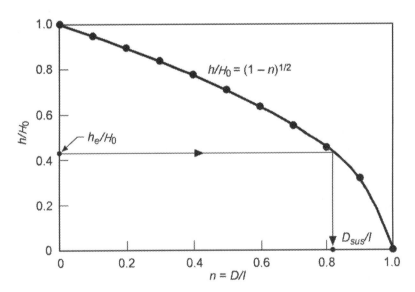

Figure 9.17 The relationship between base groundwater head and pumping volume derived from RAM2 hydrodynamic sub-model.

The hydrodynamic submodel in the RAM2 model constitutes the original Rational Analytical Model (RAM). The limitation of the RAM model is that the minimum equilibrium head h_e can only be estimated by an empirical formula. The RAM model also assumes that a sharp interface exists between the freshwater and seawater layers. Based on the shape interface assumption, a theoretical equation for calculating the minimum equilibrium head can be derived.

Salinity transport in the transition zone can be modeled using the steady-state diffusion equation:

$$\frac{\partial C}{\partial \tau} = E \frac{\partial^2 C}{\partial z^2} \qquad (9.21)$$

where τ is the travel time from the aquifer surface (or top) to the pumping well, as well as the diffusion time of salinity in the aquifer.

Suppose the salinity of seawater is C_0, and the initial and boundary conditions are as follows:

$$
\begin{aligned}
& C(0,z) = 0, z > 0; C(0,z) = C_0, z < 0 \\
& C(x, \infty) \to 0 \\
& C(x, -\infty) \to 0
\end{aligned}
\qquad (9.22)
$$

The analytical solution of Eqn. (9.21) can be obtained using Laplace transform

$$\frac{C(z,\tau)}{C_0} = \frac{1}{2}\left[1 - erf\left(\frac{z}{\sqrt{4E\tau}}\right)\right] \qquad (9.23)$$

9.2.4.2 Determination of the sustainable yield of a coastal groundwater by RAM2

For a basal aquifer, travel time τ is a constant. According to USEPA, the salinity of pumped water shall not exceed 2% of the salinity of seawater. Therefore, let $C(z,t)/C_0 = 0.02$, and Eqn. (9.23) can be written as follows:

$$0.02 = \frac{1}{2}\left[1 - erf\left(\frac{\zeta'}{\sqrt{4E\tau}}\right)\right] \qquad (9.24)$$

where ζ' is the top of the transition zone, and the sharp interface can be theoretically viewed as the midpoint of the transition zone. The application of a well-mixed ideal reactor, with consideration of the transition zone, to the basal aquifer modeling, is illustrated in Figure 9.18.

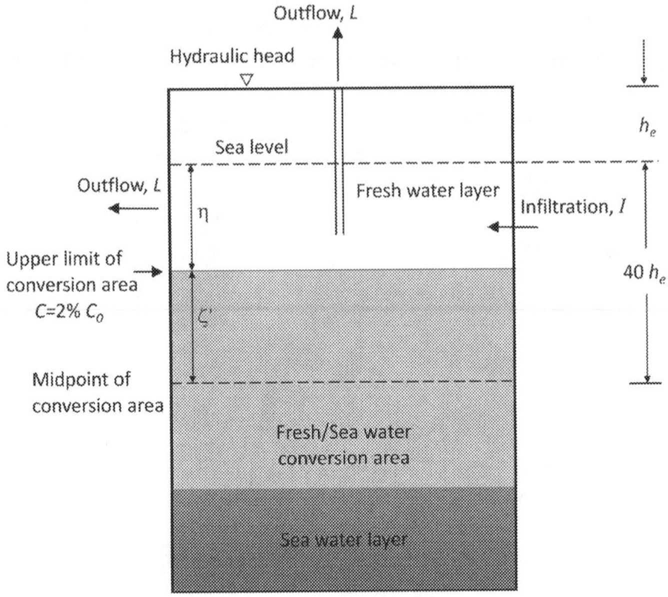

Figure 9.18 Simulation of minimum balance head of the basal aquifer.

Using Eqn. (9.24), the relationship between the top of the transition zone ζ' and the minimum equilibrium head h_e is

$$h_e = \frac{\zeta' + \eta}{40} \tag{9.25}$$

where η is the depth of the intake of the pumping well.

Example 9.3 The basal aquifer of Pearl Harbor is the largest freshwater resource in Hawaii. The area of the aquifer is 121 square miles (313 km²), and the average infiltration is 220 million gallons per day or 9.64 m³/s. The initial head H_0 before groundwater exploitation is 35 ft (10.67 m). Based on the observation records, the diffusion coefficient is estimated to be 0.776 ft²/d (0.07 m²/d), and the estimated average travel time is 38.6 years.

Problem: If the depth (η) of the intake of the pumping well is 300 ft (91.4 m), calculate the sustainable yield of the basal aquifer of Pearl Harbor.

Solution: Substituting the data into Eqn. (9.24), the following equation for ζ can be obtained:

$$0.02 = \frac{1}{2}\left[1 - erf\left(\frac{\zeta}{\sqrt{43,710}}\right)\right]$$

Using the trials and errors method, ζ is found to be 305 ft (93 m).

The minimum equilibrium head, h_e, can be computed using Eqn. (9.25):

$$h_e = \frac{\zeta + \eta}{40} = \frac{305 + 300}{40} = 15.1\,\text{ft} \ (4.6 \text{ m})$$

The dimensionless head ratio $h_e/H_o = 15.1/35.0 = 0.43$, and the dimensionless sustainable yield D_{sus}/I obtained from Figure 9.17 is 81%. The sustainable yield of the groundwater layer at the base of Pearl Harbor is $D_{sus} = 81\% \times (220) = 178$ mgd $= 7.80$ m³/s.

9.2.5   Estimation of dispersion coefficient using data of deep observation wells

For numerical modeling of coastal groundwater, an important issue is to select the correct model coefficients. We now use an example to illustrate how to estimate the dispersion coefficient using the measured salinity data from deep observation wells.

The Department of water resources of Hawaii has set up a series of observation wells in the main groundwater layers of the Hawaii islands to understand the changes in water quantity and quality, and the number of observation wells on Oahu Island is the largest (Figure 9.19). The water head and salinity can be measured at the deep groundwater observation

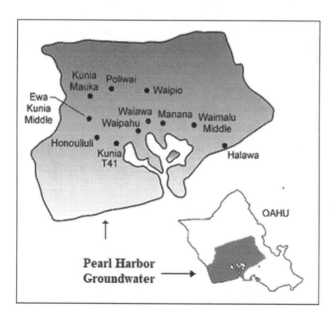

Figure 9.19 Distribution of observation wells in the groundwater layer of Pearl Harbor on Oahu Island, Hawaii.

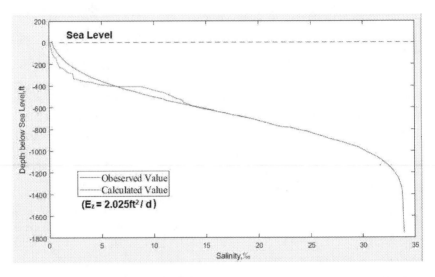

Figure 9.20 Observed and calculated results of vertical salinity distribution in the deep observation wells of Manana, Oahu Island.

stations, and the vertical profile of salinity shows the extent of the transition zone. Figure 9.20 shows the vertical distribution of salinity measured in Manana observation well.

The vertical salinity distribution in the transition zone can be modeled using the analytical equation (8.44) discussed in Section 8.5.2. The estimation of diffusion coefficient based on the measurement of natural tracer (Liu, 2007) is to compare and match the measured (observation well) and computed (Eqn. (8.44)) vertical salinity profiles by adjusting the diffusion coefficient. When the calculated vertical salinity profile agrees with the measured data, the calculated diffusion coefficient is selected as the diffusion coefficient of this aquifer.

Example 9.4   Use the vertical salinity profile (Figure 9.20) of the groundwater observation well Manana on Oahu Island to estimate the dispersion coefficient of groundwater. The effective velocity is 2.42 ft/d; and the distance from Manana observation well to the infiltration zone is 47,520 ft.

Figure 9.20 shows the vertical salinity distribution measured at the deep groundwater observation station in Manana. As this observation station is close to the coastline, there is an obvious transition zone at the bottom of the basal aquifer.

Problem: Estimate the dispersion coefficient of the basal aquifer of Pearl Harbor.

Solution: The travel time from the infiltration zone to the observation well is as follows:

$$\tau = \frac{47,520}{2.42} = 19,636 \, (d)$$

The next step is to apply Eqn. (8.47) to calculate the vertical distribution of salinity in Manana observation well. Figure 9.20 shows that when $E_Z = 2.025$ ft²/d, the calculated vertical salinity profile is closest to the observed data. Therefore, 2.025 ft²/d is the dispersion coefficient of the basal aquifer at Pearl Harbor.

9.3   MARINE WATER MODELING

9.3.1   Initial mixing of marine waste discharge

The initial mixing of marine waste discharge behavior is considered as a useful guide information to the discharge designer because the mixing process is usually sensitive to design conditions. From the environment protection point of view, when designing the waste discharge device, to minimize the risk to the local environmental, the initial mixing process of marine waste discharge must be carefully assessed.

The mixing process of the waste discharge always occurs over a wide range of spatial and temporal scales. In general, the mixing region is divided into two regions, with the region close to the discharge point being called the near field, and the region further away from the discharge point the far field. In the near field, the strong initial mixing of the waste occurs, and the mixing behavior is characterized by the waste discharge and its geometry. The spatial and temporal scales of the mixing process are relatively small in this region. As the waste transport further away from the discharge point, the ambient environment will determine the mixing and dilution characteristics of the waste. The spatial and temporal scales of the mixing process in this region are usually much larger.

In the near field, the initial mixing process is characterized by both momentum flux and buoyancy effects. The 'pure jet' forms when the momentum flux dominates in the initial mixing and the 'pure plume' forms when buoyancy flux dominates. In general, the initial mixing flow is always a combination of initial momentum flux and buoyancy flux, forming a 'buoyancy jet'. In such a case, the buoyancy jet occurs in a narrow initial zone where a violent mixing process takes place. Then, the buoyancy jet will eventually become plume-like as initial momentum is dissipated, as shown in Figure 9.21.

The field measurement, laboratory experiment, and numerical modeling are three common approaches to study the characteristics of the rapid

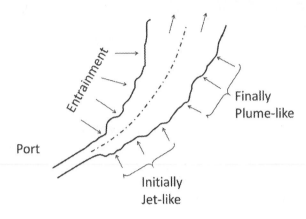

Figure 9.21 Schematic diagram of buoyancy jet.

mixing process of the waste discharge. The field measurement is always difficult and expensive due to the variation of ambient conditions and the restriction of equipment arrangement. In laboratory experiments, it is easier to control the ambient conditions and to realize the accurate and detailed measurement of the flow characteristics. The measured data can provide support for model validation and coefficient calibration. However, the laboratory scale data cannot reflect fully the mixing process in the prototype scale marine condition. Modeling approaches are widely used to predict the initial mixing process of waste discharge in the near field. By taking the waste discharge properties and the discharge configuration into account, the models can predict waste behaviors under different ambient conditions. The numerical models include length-scale models (dimensional analysis), integral models, and CFD models. A brief introduction of these models is given in the following subsections.

9.3.1.1 Length-scale models

The length-scale model is the simplest approach and can generate solutions quickly. In the model, the major influence factors on the mixing behaviors, including volume flux, buoyancy flux, and momentum flux, are considered in a non-dimensional form. For a single round jet, the fluxes in the discharge process can be expressed as follows:

The volume flux (represents the waste discharge): $Q = \frac{1}{2}\pi d_o^2 u$.

The buoyancy flux (represents the effect of gravity on the waste discharge): $B = g'Q$.

The momentum flux (represents the energy transport during the waste discharge): $M_f = \frac{1}{2}\pi d_o^2 u^2$, where u is the discharge velocity, d_o is the outlet diameter, $g' = g \dfrac{\rho_w - \rho_A}{\rho_A}$, where g is the gravity acceleration, ρ_A is the ambient density, and ρ_w is the waste density.

The mixing characteristics of a round turbulent buoyant jet in stagnant ambient water can be described by its discharge momentum flux, M_f; buoyancy flux, B; and volume flux, Q. Predictive equations were formulated in terms of two dimensionless parameters of z/L_Q and z/L_M, where z is the distance along the jet centerline, $L_Q = Q/M_f^{1/2}$, and $L_M = M_f^{3/4}/B^{1/2}$. The ratio of L_Q/L_M is defined as the jet Richardson number R_o or $L_Q/L_M = QB^{1/2}/M^{5/4}$.

For a stagnate ambient environment, the dilution rate γ_D can be expressed in terms of jet Richardson number:

$$\gamma_D = f_1(R_o) \tag{9.26}$$

where f_1 is the coefficient which can be calibrated by laboratory data.

The commercial software based on the length-scale model includes NRFIELD (Roberts et al., 1989a, 1989b, 1989c), CORMIX1 (Doneker and Jirka, 1990), and CORMIX2 (Akar and Jirka, 1991).

9.3.1.2 Integral models

In the integral model, the mass and momentum conservation equations are integrated over the cross section, assuming that there is no radial change in the velocity profile and the jet profile is axisymmetric and Gaussian. The integral model solves the governing equations along the jet axis. The waste diffusion process is modeled through the 'entrainment' approach, based mainly on the eddy viscosity concept. Fan (1967) developed a first-order integral model based on the jet entrainment approach. Besides, the second-order integral models have been developed by Wang and Law (2002), Yannopoulos (2006), and Jirka (2004).

The common commercial models of the integral type are Visual Plume (Frick, 1984) and CorJet of CORMIX (Doneker and Jirka, 2007) and VIS-JET (Lee and Cheung, 1990).

9.3.1.3 Computational fluid dynamics models

CFD models solve a general form of the Navier–Stokes equations and have fewer assumptions than the first two types of models. Direct numerical simulation (DNS) is one of the CFD models, and it solves the governing equations of flows directly, without establishing a model for turbulence field. However, it requires high spatial and temporal resolutions during the computation to obtain the flow information for all scales, and therefore, the DNS modeling is only applicable for turbulent flows with low Reynolds numbers. Many other CFD models solve the hydrodynamics and waste transport equations with the use of turbulence closure models, such as k–ε models (Worthy et al., 2001; Oliver et al., 2008) and large eddy simulation model (Basu and Mansour, 1999; Worthy, 2003; Devenish et al., 2010). Although the turbulence mixing behavior can be predicted by the numerical simulations, the models require high mesh resolution to ensure a converged and stable solution. Even with the improvement of computational efficiency, accurate CFD modeling of the near field mixing, especially for complex discharge cases, is time-consuming when compared with the length-scale and integral model approaches.

To facilitate readers to choose proper software in their respective study, a brief introduction of available software for waste discharge modeling is provided below. For rapid initial mixing process simulation in the near-field region, the most popular software packages are CORMIX, VISUAL PLUMES, and VISJET.

CORMIX (Cornell Mixing Zone Expert System) is a software system for the analysis, prediction, and design of waste discharge into the different ambient environments. The major emphasis is on the geometry and dilution characteristics of the initial mixing zone. CORMIX consists of three subsystems for simulating different discharge processes: CORMIX1 for submerged single port discharge (Doneker and Jirka, 1990), CORMIX2 for submerged multiport discharge (Akar and Jirka, 1991), and CORMIX3 for a buoyant surface jet. In addition, the CORJET model (The Cornell Buoyant Jet Integral Model), which is an integral model, is linked to the CORMIX system. The CORJET model can provide a detailed analysis of the near-field behavior of the buoyancy jet.

VISUAL PLUMES is a commercial software developed by USEPA. In this model, the waste properties, discharge configuration, and ambient conditions are considered when simulating the mixing behavior of waste discharge. The model is good at modeling the near-field region but cannot simulate the interaction between flow and boundaries. The model considers time series data and can provide time-dependent waste discharge behaviors.

VISJET is a Lagrangian model developed by the University of Hong Kong. The model has been tested against theory, laboratory experimental data, field certification, and applications. It aims to facilitate the environmental impact assessment and outfall design studies. It is able to predict the initial mixing process of waste discharge in a current.

9.3.2   Initial mixing of artificially upwelled deep ocean water

DOW at a depth of 300 m or lower is cold, nutrient-rich, and free from pathogenic bacteria. This water is an important natural resource (Takahashi, 2000). One of the fully developed technologies of DOW utilization is ocean thermal energy conversion (OTEC) which makes use of the temperature difference between cold DOW and warm surface water for energy production (Liu, 2018). Another potential DOW utilization is open ocean mariculture or 'ocean farm', which makes use of nutrient-rich DOW to enrich the open oceans and thus to increase the yield of fish and other marine products (Liu, 2018).

The open oceans in areas other than where natural upwelling occurs are characterized as 'wet deserts' – the average net primary productivity (NPP) for the open oceans is approximately 80 g carbon per square meter per year, compared to an average NPP of over 800 g carbon per square meter per year in tropical forests (Miller, 1985). DOW represents an ideal source of nutrients for photosynthesis and aquaculture feedstock production – upwelling regions account for only 0.1% of the world's ocean surface but yield roughly 44% of the fish catch (Redfield et al., 1963).

Land-based aquaculture, using DOW pumped from the ocean depths into man-made ponds and enclosures for growing algae (kelp, Gracilaria, and Spirulina) and primary herbivores (abalone), has been in existence in the USA and Japan for more than three decades (Daniel, 1984; Grove et al., 1989; Nakashima et al., 1989). However, land-based aquaculture is small in scale. Providing adequate food supply for the world's increasing population requires a large-scale development of open ocean mariculture. Currently, principal research needs for this development include the pumping cost-effectively of DOW to ocean surface or artificial upwelling, and the containing of DOW effluent plume within the biologically productive area of the open ocean without significant dilution. To meet these needs, a two-phase research project on artificial upwelling and mixing (AUMIX) was conducted by researchers at the University of Hawaii and Zhejiang University (Liu, 1999; Fan et al., 2015) (Figure 9.22). Phonetically, AUMIX in Chinese is '奥祕', which means 'deep mystery'.

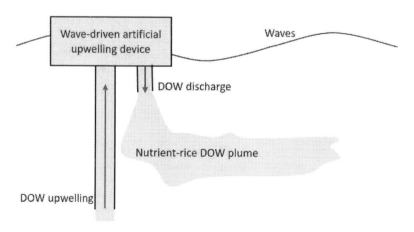

Figure 9.22 Schematic diagram of artificial upwelling and mixing.

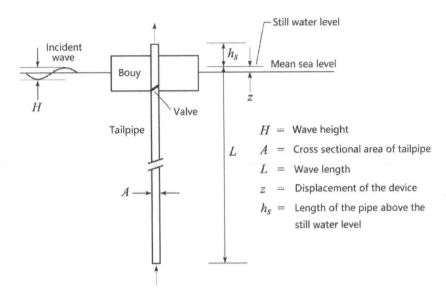

Figure 9.23 Wave-driven artificial upwelling.

The objective of the first phase of AUMIX research was to develop a wave pump, which brings DOW cost-effectively to the surface (Liu and Jin, 1995; Liu et al., 1999). The objective of the second phase of AUMIX research was to identify an optimal DOW discharge design to establish a desirable DOW plume in the open oceans (Liu et al., 2003).

A wave-driven artificial upwelling device (Figure 9.23) to convert wave power to the kinetic energy of upwelled DOW was developed and tested (Chen et al., 1995). As the wave crest approaches, the flow-controlling valve of the device is closed and the water column inside the device rises together with the device. As the wave descends, the valve is opened and the water column continues its upward movement due to inertia. Therefore, when a device moves up and down in the ambient waves, the water column inside the device keeps moving upward and brings the DOW to the surface. The efficiency of the device depends on the relative movement of the device and the water column inside. This was investigated by laboratory and field experiments as well as mathematical modeling (Liu, 1999).

Simple mathematical models of wave-driven artificial upwelling were formulated by ignoring the relative movement of the device and the water column inside the device or by assuming that the device follows exactly the movement of ambient ocean waves (Issacs et al., 1976; Vershinskiy et al., 1987; Liu et al., 1999). Field measured upwelling flow rates were higher than that calculated by simple mathematical modeling analysis (Vershinskiy et al., 1987).

More detailed analysis is conducted by considering the relative movement of the device and the water column, which can be represented by three parameters: added mass, damping coefficient, and exciting force (Liu and Jin, 1995; Liu et al., 1999). The added mass and damping coefficient indicate the extent of resistance and are functions of the device design, such as the dimensions of the device and the tailpipe length. The exciting force indicates the magnitude of an external force acting on the device and is a function of incident waves. A series of hydraulic experiments was conducted in a wave basin at the University of Hawaii's Oceanographic Engineering Laboratory to measure these three parameters under various design conditions (Figure 9.24(a)).

Field experiments were conducted on July 20, 1996, and on August 16, 1998, about 1 mile off the southern coast of Oahu, Hawaii (Figure 9.24(b)). The wave-driven artificial upwelling device used was a 1/10 model of a full-size device (Liu, 1999). In the 1996 experiment, a device with an 80-ft

(a)

(b)

Figure 9.24 Testing the wave-driven artificial upwelling device (a) at the University of Hawaii Oceanographic Engineering Laboratory and (b) in the open ocean of Oahu, Hawaii.

tailpipe was deployed in ocean waves that were 1 to 3 ft high and had a period of about 8 s. The experimental data showed that an upwelling flow of 4×10^{-3} m³/s (0.14 ft³/s) was produced. In the 1998 experiment, devices with 40-ft and 100-ft tailpipes were deployed, and more detailed measurements of the upwelling flow rate and the mixing characteristics were conducted. The average flow rate was 2×10^{-3} m³/s (0.07 ft³/s) for the 40-ft tailpipe device and 2.3×10^{-3} m³/s (0.08 ft³/s) for the 100-ft tailpipe device. These results are consistent with those calculated by the mathematical modeling analysis as discussed later (Liu and Jin, 1995).

When the valve is closed, the velocity of the water column relative to the upwelling device is zero or $U = 0$.

Under this condition, the equation of motion of the device takes the following form:

$$\left(M_f' + m_w\right)\ddot{z} = -M_a\ddot{z} - f_d\dot{z} - \beta|\dot{z}|\dot{z} - \rho g A_d z + F_e \qquad (9.26)$$

where m_w is the mass of the water in the pipe, z is the displacement of the heave of the buoy above the still water line, $\dot{z}$ is the velocity, and $\ddot{z}$ is the acceleration; A_d is the device's cross-sectional area at the still water line; M_f' is the mass of the floating system; M_a is the added mass; f_d is the damping coefficient; and F_e is the wave exciting force in the vertical direction.

When the valve is opened, the relative acceleration of the water column to the device can be determined by

$$\dot{U} + \ddot{z} + \frac{z + h_s}{L + h_s}g = 0 \qquad (9.27)$$

where h_s is the length of the pipe above the still water level

The equation of motion of the device then takes the form of

$$M_f\ddot{z} = -M_a\ddot{z} - f_d\dot{z} - \beta|\dot{z}|\dot{z} - \beta'U^2 - \rho g A_d z + F_e \qquad (9.28)$$

Equations (9.27) and (9.29) are similar except that in Eqn. (9.27) both the mass of the floating device and the mass of the water in the pipe are considered, but the relative movement of the water inside the pipe is not. On the other hand, in Eqn. (9.29), the mass of water in the pipe is not included but the viscous effect is.

According to the three-dimensional linear wave theory, values of M_a, f_d, and F_e can be determined by the integration of velocity potentials over wetted surfaces (Faltinsen and Michelsen, 1974). The set of modeling equations – along with known M_a, f_d, and F_e – constitutes a general mathematical model of wave-driven artificial upwelling. A computer program was prepared to solve modeling equations numerically using the Runge–Kutta method (Liu and Jin, 1995). This model was used to evaluate the performance of a wave-driven artificial upwelling device in actual field conditions. As part of the mathematical modeling analysis, a time series representing Hawaii random waves was developed on the basis of available field data and the general form of the Bretschneider (1969) spectrum.

In this mathematical modeling analysis, the artificial upwelling device has a 4.0-m-diameter buoy and a 300-m-long by 1.2-m-diameter tailpipe, and the height and period of the regular waves (1.90 m and 12.10 s, respectively) are the same as the significant wave height and period of random Hawaii waves. Modeling results indicate that this device can generate an upwelling flow of 0.95 m³/s in Hawaii random waves and an upwelling flow of 0.45 m³/s in regular waves (Liu and Jin, 1995). An upwelling rate of 2×10^{-3} m³/s (0.07 ft³/s) was measured in the 1998 field experiments using a 1/10 model. For a full-size prototype, the upwelling flow rate would be about $2 \times 10^{-3} \times \sqrt{10^5} = 0.63$ m³/s (22.14 ft³/s), which is very close to the results derived by mathematical modeling.

As the cold DOW effluent is heavier than the surface ocean, it enters the ambient ocean as a negatively buoyant jet (Figure 9.25). The mixing process in this near-field zone is rapid because the velocity difference and the density difference between the cold DOW effluent and the warm ocean surface are large. The equilibrium depth of the plume is reached when this density difference approaches zero and the velocity difference between the discharge and the ambient ocean becomes small. The diluted plume is then carried away by ocean currents and eddies to form a transition zone where further dilution is relatively small compared to the initial dilution because turbulence mixing is mainly caused by the velocity difference between the plume and the ambient water (Figure 9.25). Beyond the transition zone,

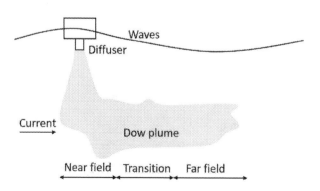

Figure 9.25 Near-field mixing of a DOW plume.

turbulence diffusion becomes the dominant plume mixing mechanism, and it is defined as the far-field zone (Figure 9.25). The objective of the AUMIX research is to create and maintain a nutrient-rich DOW plume in the open ocean by proper engineering design of effluent diffuser; therefore, only the DOW plume mixing in the near field where most of the effluent dilution and descending takes is investigated in this section.

Most of the past studies on near-field mixing were conducted in relation to the ocean disposal of wastewater and power generation cooling water (Muellenhoff et al., 1985). The principal design objective was to achieve a higher degree of effluent dilution, contrary to the objective of an AUMIX system. Also, municipal effluent is less dense than ocean water so that it behaves as a buoyant jet moving upward toward the ocean surface. However, the basic principles of the near-field mixing of municipal wastewater effluent as a buoyant jet and the near-field mixing of DOW effluent as a negatively buoyant jet are essentially the same.

In a preliminary study, simple predictive equations can be used. The simple predictive equations provide useful 'order of magnitude' estimates of near-field mixing.

A turbulent buoyant jet has both jet and plume characteristics. Near the orifice, a turbulent jet behaves like a simple jet. A simple predictive equation of the mean dilution D_{sm} of a simple jet was formulated by dimensional analysis and was tested with experimental data (Fischer et al., 1979),

$$D_{sm} = 0.25 \frac{z}{l_Q} \tag{9.29}$$

Far away from the orifice, a turbulent jet becomes a buoyancy-dominated jet, and the mean dilution, D_{sm}, can be calculated by a simple relationship,

$$D_{sm} = 0.1 \left(\frac{z}{l_Q} \right)^{5/3} \left(\frac{R_0}{R_P} \right)^{2/3} \tag{9.30}$$

where R_P is the plume Richardson number, which takes a constant value of about 0.557.

The mixing characteristics and trajectory of a buoyant jet under ambient crossflow were established by Wright (1977) in terms of two characteristic length scales: $Z_M = M^{1/2}/U$ and $Z_B = B/U^3$, where U is the ambient current velocity.

The jet mixing process is often affected by surface waves; therefore, wave effects on turbulent jet mixing have been studied extensively (Ismail and Wiegel, 1983; Chin, 1987; Chyan and Hwung, 1993; Hwang et al., 1996; Xu et al., 2017). Chin (1987) used the maximum horizontal wave-induced velocity at the discharge point u_{max} to evaluate wave effects on jet mixing to get

$$\frac{D_S}{D_{S_0}} = 1 + 6.15\left(\frac{l_Q}{Z_M}\right) \tag{9.31}$$

where D_S and D_{S_0} are the surface dilution with and without wave effects, respectively, and $Z_M = M^{1/2}/u_{max}$ is a characteristic length scale indicating the relative importance of waves on initial mixing.

As part of the AUMIX research, Liu and Lin (1998) investigated wave effects on the near-field mixing of a buoyant jet by conducting hydraulic modeling experiments in a 12 m × 1.2 m × 0.9 m glass wave tank. The diameter of the injection nozzle orifice used was 0.45 mm. Jet discharges larger than 14 cm³/s were used in the experiments to ensure turbulent flow (Figure 9.26).

The results of the hydraulic experiments indicate that wave characteristic r_0 as well as u_{max} must be used in evaluating the effect of waves on jet mixing. The variable r_0 is the maximum wave-induced water particle displacement at the discharge port. A functional relation for predicting trajectory and dilution of a buoyant jet under ambient crossflow and ocean wave was established (Liu and Lin, 1998) as

$$D_S = D_{Sc} + \frac{r_0 z - Z_0}{l_Q \times Z_m} \tag{9.32}$$

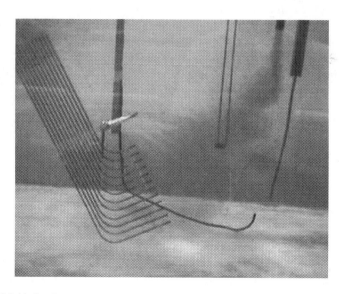

Figure 9.26 Hydraulic experiment of near-field DOW mixing under the actions of waves and cross flow.

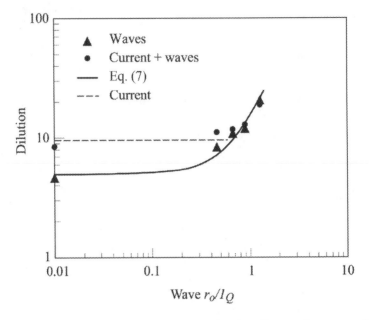

Figure 9.27 Dilution of a turbulent buoyant jet under the influence of current and wave actions.

where D_{Sc} is the dilution of a buoyant jet along its centerline without wave actions, z is the vertical downward distance measured from the still water level, and z_0 is the location of the discharge point below the still water level.

Figure 9.27 shows the results of a typical experimental run conducted under the following discharge and ambient conditions: ratio of current velocity (U) and jet discharge velocity (U_0), $R = 0.11$; jet densiometric Froude number $F_d = 8.3$; $z/l_0 = 20$; and $z/Z_B = 2.23$.
Equation (9.33) was formulated on the basis of experimental data collected in shallow to intermediate-water waves. In deep-water waves, the wave-induced particle movement declines exponentially with water depth (Dean and Dalrymple, 1991) and Eqn. (9.33) can be modified as follows:

$$D_S = D_{Sc} + \frac{r_0}{l_Q} \frac{\frac{1}{k}\left(e^{-kZ_0} - e^{-kZ}\right)}{Z_m} \tag{9.33}$$

where k is the wavenumber.

More detailed mathematical modeling of near-field DOW plume mixing can be conducted with a modified UM model. The UM model, developed by the USEPA (Baumgartner et al., 1994) was formulated on the basis of mass and momentum conservation equations and the Taylor entrainment hypothesis. In the UM model, total entrainment is the sum of the Taylor entrainment and forced entrainment. Taylor entrainment is caused by the internal turbulence of buoyant jets, whereas forced entrainment is caused by currents.

The model is built on the basis of equations of the conservation of mass, momentum, and heat energy. These equations are solved numerically by following a Lagrangian approach, which takes a buoyant jet as serial elements following the same trajectory.

In previous studies, wave effects on the initial dilution of a buoyant jet were included in the mathematical model by extending the forced entrainment function (Chin, 1987; Hwang et al., 1996). Following this approach, the characteristic jet velocity in the entrainment function was defined as the relative velocity of the jet and the instantaneous wave-induced velocity. Since the wave velocity is time-dependent, the entrainment function would also be time-dependent. However, it is inappropriate to use a time-dependent entrainment function in the UM model, which is a steady-state mathematical model. Furthermore, including the wave effects in the forced entrainment function is inconsistent with the linear wave theory, which suggests that waves are not able to transport mass. The results of the hydraulic experiments conducted by this study, however, indicate that vertical buoyant jets extend laterally faster when surface waves are present. These results suggest that wave effects can be included by extending the Taylor entrainment, which is similar to the way the buoyancy effects are included (Fox, 1970).

Taking into consideration wave and buoyancy effects, the Taylor entrainment function for round buoyant jets takes the following form (Liu et al., 2003):

$$\alpha_\tau = \alpha_1 + \frac{\alpha_2}{F_r} + \alpha_w \tag{9.34}$$

In Eqn. (9.34), the total Taylor entrainment coefficient is the sum of three coefficients: α_1, which is due to the internal jet turbulence; α_2/F_r, which accounts for the buoyancy effects; and α_w, which accounts for the effects of waves. The third coefficient can be expressed as follows:

$$\alpha_w = \lambda_w \frac{d_r}{R_e} \tag{9.35}$$

where λ_w is a proportionality coefficient, d_r is the local effective radius of wave-driven tracer particle displacement that is calculated from the local horizontal and vertical displacements, and R_e is the local Reynolds number.

By fitting experimental data produced by the hydraulic modeling experiment with the UM modeling results, the coefficient λ_w was found to be 150

(Liu et al., 2003). Thus, for round buoyant jets, the total Taylor entrainment, including the effects of buoyancy and waves, can be expressed as follows:

$$E_{total} = \left(\alpha_1 + \frac{\alpha_2}{Fr} + \alpha_3 \frac{d_r}{l_Q R_e} \right) 2\pi b u_m \qquad (9.36)$$

For the modified UM model, $\alpha_1 = 0.13$, $\alpha_2 = 0.97$, and $\alpha_3 = 150$.

Equation (9.37) was used to investigate the mixing of DOW effluent brought up by an artificial upwelling device. The following environmental conditions used in the artificial upwelling study were adopted (Liu and Jin, 1995): discharge nozzle diameter = 1.2 m, effluent flow rate = 0.95 m³/s, wave height = 1.9 m, and wave period = 12 s. A typical tropical ambient temperature profile (Paddock and Ditmas, 1983) was applied. The effluent temperature was assumed to be 15°C. The salinity of both the ambient ocean water and the DOW effluent was 3.5%. The investigation was conducted by positioning the injection nozzle at depths of 0, 5, 10, 20, 30, 40, and 50 m.

The calculated effluent dilution (Figure 9.28(a)) shows that wave effects on jet mixing are insignificant when the injection nozzle is positioned 30 m or more below the ocean surface. Only the density difference between the effluent plume and its ambient ocean water is considered in calculating descending depth (Figure 9.28(b)), as the surface waves have smaller effects on the descending depth than on the initial dilution.

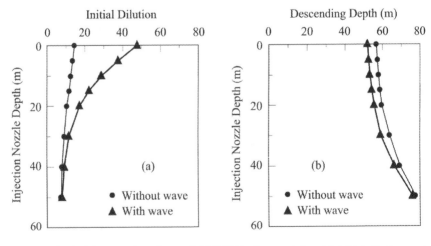

Figure 9.28 Modeling results of near-field DOW plume mixing: (a) initial dilution and (b) descending depth.

The effects of currents and waves on the dilution of DOW effluent discharged at the ocean surface are not accumulative. In areas of strong currents and weak waves ($r_0/l_Q < 1.0$), the wave effect on the dilution is negligible. On the other hand, in areas of small crossflow currents and strong waves ($r_0/l_Q > 1.0$), the current effect on dilution is negligible. Hence, a nutrient-rich DOW plume can be established and sustained in the open ocean by wave-driven artificial upwelling and by proper discharge of DOW to the upper layer of the open ocean.

The formation of a nutrient-rich region in the open ocean requires controlling the DOW plume such that the dilution of the nutrients in the plume is small and the descending depth is within the surface trophic zone where primary productivity takes place.

The modified UM model was used to investigate DOW plume dilution and descending depth. The investigation involved discharging upwelled DOW into the open ocean at a flow rate of 0.95 m^3/s, which corresponds to the pumping rate produced by a prototype wave-driven artificial upwelling device (Liu and Jin, 1995). Modeling analysis considered the discharge of DOW with one or more diffuser ports, at different depths and different discharge angles. Also considered was the ambient density stratification, an important factor affecting the initial dilution and the equilibrium depth of the DOW plume. In this study, we need a typical density profile in the tropical ocean, which shows a constant density over a top depth of 50 m. This profile was obtained from a technical report by Paddok and Ditmas (1983).

First, wave effects on DOW plume mixing were investigated. Modeling results showed that wave effects on the initial dilution of a DOW plume are insignificant relative to buoyancy effects. Under the condition of deep-water waves, which is the typical condition for a DOW discharge, particle displacement decreases exponentially with depth as well as the wave-induced entrainment coefficient. In a particular case of a vertical discharge of 0.95 m^3/s DOW through a single discharge port of 1 m in diameter, the initial momentum is relatively strong, with the Reynolds number equal to 1.21×10^6. At a wave height of 1.9 m, a wave period of 12 s, and a water depth of 800 m, the maximum relative radius, dw, is 1.34 m at the surface. Thus, the wave-induced entrainment coefficient, aw, is very small with an initial value of 1.88×10^{-4}. The wave effect is insignificant regardless of the depth of DOW effluent discharge.

The modeling results of near-field mixing of the DOW effluent from two diffuser ports pointing vertically downward are shown in Figure 9.28. The equilibrium depth is the depth a buoyant jet reaches when differences of density and velocity between a jet and ambient ocean water disappear. Apparently, without any effluent control (or the base case), the DOW effluent would be too diluted and hence would travel below the euphotic zone. Control of DOW effluent buoyancy can be achieved by raising the temperature of the effluent. Figure 9.28 shows that by reducing the effluent buoyancy by 75% and by locating the diffuser at a depth of 40 m, a DOW effluent with 10 to 1 dilution can be maintained. Under these conditions, a nutrient-rich plume can be maintained at a depth of 58 m.

More recently, a model of near-field mixing of DOW Effluent Plume was formulated on the basis of the Reynolds-averaged Navier–Stokes equations and was used to investigate the feasibility of the creation of a nutrient-rich plume in the South China Sea (Fan et al., 2015). Modeling results indicate that a nutrient-rich DOW plume can be created and sustained in the South China Sea by discharging the DOW at an optimal depth.

EXERCISES

1. In order to prevent the expansion of the pollution scope of the underground water layer, the area polluted by pesticides and the area not polluted shall be separated by a vertical clay wall. The width w of the clay wall is 3 m (see Figure 9.29). The hydraulic conductivity of clay is 5×10^{-8} m/s, and the porosity is 0.4. The concentration of pesticides in polluted groundwater is 10 ppb, and the water head H_0 is 12 m, and the water head H of unpolluted groundwater is 5 m.

   (a) Calculate the time required for the pollutant to pass through the clay wall, assuming that the pesticide has no adsorption in the clay.
   (b) Calculate the concentration of pesticides entering the unpolluted aquifer. The first-order decomposition of pesticide is carried out in clay, and the reaction coefficient k was 0.01 d^{-1}.

2. If the pesticide has an adsorption effect in the clay, the adsorption coefficient $K_D = 1.8 \times 10^{-3}$ m^3/kg, and the density of clay (bulk density) = 1250 kg/m^3. Repeat the question above.

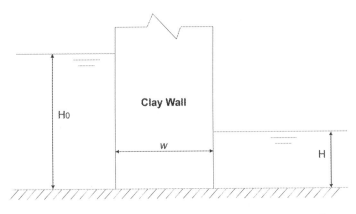

Figure 9.29 Illustration of the pollution of the underground water layer (Exercise 1).

3. A long-term leakage organic matter (Benzene) from an underground oil storage tank enters a one-dimensional underground water layer with a cross-sectional area of 10 m². The hydraulic conductivity of the water layer is 2.15 m/d, the effective porosity is 0.1, the hydraulic slope is 0.04, and the diffusion coefficient is 7.5 m. If the initial concentration is 1000 g/m³, calculate how long it will take for the concentration of benzene to reach 100 g/m³ at 750 m downstream. Suppose that benzene is a conservative compound.

4. Another underground oil storage tank accidentally released 1 kg (1000 g) of non-conservative compounds into the one-dimensional groundwater layer in the previous question. The first-order attenuation coefficient of the compound is 0.021 y^{-1}.

   (a) Calculate the initial concentration if the compounds are completely mixed in the cross-sectional area immediately after entering the groundwater layer.
   (b) Calculate the concentration at 100 m downstream after 90 days.

5. List and discuss the assumptions introduced in deriving the RAM2 model.

REFERENCES

Akar, P.J. and Jirka, G.H. (1991). *CORMIX2: An Expert System for Hydrodynamic Mixing Zone Analysis of Conventional and Toxic Submerged Multiport Diffuser Discharges*. U.S. Environmental Protection Agency (EPA), Office of Research and Development.

Alley, W.M. and Leake, S.A. (2004). The journey from safe yield to sustainability. *Groundwater*, 42(1), pp. 12–16.

Ayers, J.F. and Vacher, H.L. (1983). A numerical model describing unsteady flow in a fresh water lens. *Water Resources Bulletin*, 19(5), pp. 785–792.

Basu, A.J. and Mansour, N.N. (1999). *Large Eddy Simulation of a Forced Round Turbulent Buoyant Plume in Neutral Surroundings*. Annual Research Briefs, Center for Turbulence Research, pp. 239–248.

Baumgartner, D.J., Frick, W.E. and Roberts, P.J.W. (1994). *Dilution Models for Effluent Discharges, EPA/600/R-94/086*. U.S. Environmental Protection Agency.

Bear, J. (1979). *Hydraulics of Groundwater*. McGraw-Hill Inc., New York.

Bretschneider, C.L. (1969). Wave forecasting. Chapter 11 in *Handbook of Ocean and Under Water Engineering*. McGraw-Hill Book Co., New York.

Chen, H., Liu, C.C.K. and Guo, F. (1995). Hydraulic modeling of wave-driven artificial upwelling. *Journal of Marine Environmental Engineering*, 1, pp. 263–277.

Chin, D.A. (1987). Influence of surface waves on outfall dilution. *Journal of Hydraulic Engineering*, ASCE, 113(8), pp. 1006–1017.

Chyan, J.M. and Hwung, H.H. (1993). On the interaction of a turbulent jet with waves. *Journal of Hydraulic Research*, 31(6), pp. 791–810.

Contractor, D.N. (1983). Numerical modeling of saltwater intrusion in the northern guam lens 1. *JAWRA Journal of the American Water Resources Association*, 19(5), pp. 745–751.

Custodio, E. and Bruggeman, G.A. (1987). *Groundwater Problems in Coastal Areas*. Studies and Reports in Hydrology (UNESCO).

Daniel, T.H. (1984). OTEC and cold water aquaculture research at the natural energy laboratory of Hawaii. *Proceedings of PACON'84, MRM 2/47–52*. Honolulu, HI.

Dean, R.G. and Dalrymple, R.A. (1991). *Water Wave Mechanics for Engineers and Scientists*. World Scientific, Singapore.

Deltares. (2015). *Sinking Cities: An Integrated Approach towards Solutions*. Available at: https://www.deltares.nl/app/uploads/2015/09/Sinking-cities.pdf [Accessed on 25 June 2021].

Devenish, B.J., Rooney, G.G. and Thomson, D.J. (2010). Large-eddy simulation of a buoyant plume in uniform and stably stratified environments. *Journal of Fluid Mechanics*, 652, p. 75.

Doneker, R.L. and Jirka, G.H. (1990). *Expert System for Hydrodynamic Mixing Zone Analysis of Conventional and Toxic Submerged Single Port Discharges (CORMIX1)*. Cornell University Ithaca United States.

Doneker, R.L. and Jirka, G.H. (2007). *CORMIX User Manual: A Hydrodynamic Mixing Zone Model and Decision Support System for Pollutant Discharges into Surface Waters*. MixZon Inc., Portland, OR.

Engineers, T. (2005). Development of a groundwater management model, Honolulu Area of the Southern Oahu groundwater system. *Prepared for the City and County of Honolulu Board of Water Supply*. Honolulu, HI.

Essaid, H.I. (1990). A multilayered sharp interface model of coupled freshwater and saltwater flow in coastal systems: Model development and application. *Water Resources Research*, 26(7), pp. 1431–1454.

Faltinsen, O.M. and Michelsen, F.C. (1974). Motions of large structures in Waves at Zero Froude Number. *Proceedings of the International Symposium on the Dynamics of Marine Vehicles and Structures in Waves*. University College, London, pp. 91–106.

Fan, L.N. (1967). *Turbulent Buoyant Jets into Stratified or Flowing Ambient Fluids*, PhD. California Institute of Technology, Pasadena.

Fan, W., Pan, Y., Liu, C.C.K., Wiltshire, J.C. and Chen, C.A. (2015). Hydrodynamic design of deep ocean water discharge for the creation of a nutrient-rich plume in the South China Sea. *Journal of Ocean Engineering*, 108, pp. 356–368.

Fischer, H.B., List, E.J., Koh, R.C.Y., Imberger, J. and Brooks, N.H. (1979). *Mixing in Inland and Coastal Waters*. Academic Press, New York.

Fox, D.G. (1970). Forced plume in a stratified fluid. *Journal of Geophysical Research*, 75(33), pp. 6818–6835.

Frick, W.E. (1984). Non-empirical closure of the plume equations. *Atmos Environ*, 18(4), pp. 653–662.

Frind, E.O. (1982). Simulation of long-term transient density-dependent transport in groundwater. *Advances in Water Resources*, 5(2), pp. 73–88.

Gingerich, S.B. and Voss, C.I. (2005). Three-dimensional variable-density flow simulation of a coastal aquifer in Southern Oahu, Hawaii, USA. *Journal of Hydrogeology*, 13, pp. 436–450.

Grove, R.S., Sonu, C.J. and Nakamura, M. (1989). Recent Japanese trends in fishing reef design and planning. *Journal of Marine Science*, 44, pp. 984–996.

Guswa, J.H. and LeBlanc, D.R. (1980). *Digital Models of Ground-water Flow in the Cape Cod Aquifer System, Massachusetts.* US Government Printing Office, p. 112.

Henry, H.R. (1964). Effects of dispersion on salt encroachment in coastal aquifers, in "Seawater in Coastal Aquifers". *US Geological Survey, Water Supply Paper,* 1613, pp. C70–C80.

Huisman, L. (1972). *Groundwater Recovery.* Winchester Press, New York.

Hwang, R.R., Yang, W.C. and Chiang, T.P. (1996). Effect of surface waves on a buoyant jet. *Journal of Marine Environmental Engineering,* 3, pp. 63–84.

IPCC. (2013). *Climate Change 2013: The Physical Science Basis. Contribution of Working Group I to the Fifth Assessment Report of the Intergovernmental Panel on Climate Change.* Cambridge University Press, Cambridge, UK and New York, NY.

Ismail, N.M. and Wiegel, R.L. (1983). Opposing wave effects on momentum jets spreading rate. *Waterway, Port, Coastal, and Ocean Engineering, American Society of Civil Engineers (ASCE),* 109(4), pp. 465–486.

Issacs, J.D., Castel, D. and Wick, G.L. (1976). Utilization of the energy in ocean waves. *Journal of Ocean Engineering,* 3, pp. 175–187.

Jirka, G.H. (2004). Integral model for turbulent buoyant jets in unbounded stratified flows. part I: Single round jet. *Environmental Fluid Mechanics,* 4, pp. 1–56.

Lau, L.S. and Mink, J.F. (2006). *Hydrology of the Hawaiian Islands.* University of Hawaii Press, Honolulu, Hawaii.

Lee, J.H. and Cheung, V. (1990). Generalized Lagrangian model for buoyant jets in current. *Journal of Environmental Engineering,* 116(6), pp. 1085–1106.

Liu, C.C.K. (1999). Research on artificial upwelling and mixing at the University of Hawaii at Manoa. *IOA Newsletter. International OTEC/DOWA Association,* 10(4), pp. 1–8.

Liu, C.C.K. (2007). RAM2 modeling and the determination of sustainable yield of Hawaii Basal Aquifer. *Project Report PR-2009-06,* Water Resources Research Center. University of Hawaii, 81p.

Liu, C.C.K. (2018). Ocean thermal energy conversion and open ocean mariculture: The prospect of Mainland-Taiwan collaborative research and development. *Sustainable Environment Research,* 28, pp. 267–273.

Liu, C.C.K. and Dai, J. (2012). Seawater intrusion and sustainable yield of basal aquifers Filtering of dissolved oxygen data in stream water quality analysis. *Journal of the American Association of Water Resources (AWRA),* 48(5), pp. 861–870.

Liu, C.C.K., Dai, J., Lin., H. and Guo, F. (1999). Hydrodynamic performance of wave-drive artificial upwelling. *Journal of Engineering Mechanics, ASCE,* 125(7), pp. 728–732.

Liu, C.C.K., Ewart, C. and Huang, Q. (1991). Response of a basal water-body to forced draft. In Peters, J. (ed.), *ASCE Book: Ground Water in the Pacific Rim Countries.* American Society of Civil Engineers (ASCE), New York, pp. 36–42.

Liu, C.C.K., Green, R.E., Lee, C.C. and Williams, M.K. (1983). Modeling analysis of pesticide DBCP transport and transformation in soils of Kunia Area in Central Oahu, Hawaii. *Completion Report Submitted to US Environmental Protection Agency, Pacific Biomedical Research Center.* University of Hawaii, 51p.

Liu, C.C.K. and Jin, Q. (1995). Artificial upwelling in regular and random waves. *Journal of Ocean Engineering*, 22(4), pp. 337–350.

Liu, C.C.K. and Lin, H. (1998). Discharge and mixing of artificially upwelled deep ocean water. *Environmental Hydraulics: Proceedings of the Second International Symposium on Environmental Hydraulics*. A.A. Balkema, Rotterdam, Netherlands, pp. 149–154.

Liu, C.C.K., Moravcik, P., Fernandes, K., Card, B.J. and Lee, T. (2012). Survey and modelling analysis of HDOT MS4 highway storm runoff on Oahu, Hawaii. *WRRC Project Report PR-2012-01*. Water Resources Research Center, University of Hawaii at Manoa, Honolulu, HI.

Liu, C.C.K., Sou, I.M. and Lin, H. (2003). Artificial upwelling and near-field mixing of a nutrient-rich deep-ocean water plume. *Journal of Marine Environmental Engineering*, 7(1), pp. 1–14.

Maimone, M. (2004). Defining and managing sustainable yield. *Ground Water*, 42(6), pp. 809–814.

Miller, G.T. (1985). *Living in the Environment: An Introduction to Environmental Science* (4th ed.). Wadsworth, Belmont, CA.

Mink, J.F. (1980). *State of the Groundwater Resources of Southern Oahu*. Board of Water Supply, City and County of Honolulu, Honolulu, HI.

Mink, J.F. (1981). Determination of sustainable yields in basal aquifers of Hawaii. In *Groundwater in Hawaii: A Century of Progress*. Water Resources Research Center, University of Hawaii at Manoa, Honolulu, HI, pp. 101–116.

Muellenhoff, W.P., Soldate, A.M., Baumgartner, D.J., Schuldt, M.D., Davis, L.R. and Frick, W.E. (1985). *Initial Mixing Characteristics of Municipal Ocean Discharge, Volume I – Procedures and Applications*, EPA/600/3-85/073a. USEPA, Newport, Oregon.

Nakashima, T. Toyata, T. and Fujita, T. (1989). Research and development of mariculture systems using deep ocean water. In *Proceedings Int. Workshop on Artificial Upwelling and Mixing in Coastal Waters*. Center for Ocean Resources Yechnology, University of Hawaii, Honolulu, HI, pp. 68–87.

Noda, E.K. (1992). *AlaWai Canal Improvement Project Feasibility Report*. Nodal and Associates, Inc., Honolulu, HI.

Oliver, C.J., Davidson, M.J. and Nokes, R.I. (2008). k-ε Predictions of the initial mixing of desalination discharges. *Environmental Fluid Mechanics*, 8(5), pp. 617–625.

Paddock, R.A. and Ditmas, J.D. (1983). *Initial Screening of License Application for Ocean Thermal Energy Conversation (OTEC) Plants with Regard to Their Interaction with the Environment*. ANL/OTEC-EV-2, Ocean Thermal Conversation Program, Argonne National Laboratory.

Redfield, A.C., Ketchum, B.H. and Richards, F.A. (1963). The influence of organisms on the composition of seawater. *The Sea*, 2, pp. 26–77.

Roberts, P.J.W., Snyder, W.H. and Baumgartner, D.J. (1989a). Ocean outfalls. I: Submerged wastefield formation. *Journal of Hydraulic Engineering*, 115(1), pp. 1–25.

Roberts, P.J.W., Snyder, W.H. and Baumgartner, D.J. (1989b). Ocean outfalls. II: Spatial evolution of submerged wastefield. *Journal of Hydraulic Engineering*, 115(1), pp. 26–48.

Roberts, P.J.W., Snyder, W.H. and Baumgartner, D.J. (1989c). Ocean outfalls. III: Effect of diffuser design on submerged wastefield. *Journal of Hydraulic Engineering, ASCE*, 115(1), pp. 49–70.

Segol, G. and Pinder, G.F. (1976). Transient simulation of saltwater intrusion in southeastern Florida. *Water Resources Research*, 12(1), pp. 65–70.

Segol, G., Pinder, G.F. and Gray, W.G. (1975). A Galerkin-finite element technique for calculating the transient position of the saltwater front. *Water Resources Research*, 11(2), pp. 343–347.

Seward, P., Xu, Y. and Brendock, L. (2006). Sustainable groundwater use, the capture principle, and adaptive management. *Water SA*, 32(4), pp. 473–482.

Sophocleous, M. (1997). Managing water resources systems: Why "safe yield" is not sustainable. *Ground Water*, 35(4), p. 561.

Syvitski, J.P., Kettner, A.J., Overeem, I., Hutton, E.W., Hannon, M.T., Brakenridge, G.R. and Nicholls, R.J. (2009). Sinking deltas due to human activities. *Nature Geoscience*, 2(10), pp. 681–686.

Takahashi, M. (2000). *DOW: Deep Ocean Water as Our Next Natural Resource*. Terra Scientific Publishing Company, Tokyo, Japan.

U.S. EPA. (1977). Water quality assessment: A screening method for nondesignated 208 areas. *EPA-600/9-77/023, U.S. Environmental Protection Agency Office of Research and Development*. Environmental Research Laboratory, Athens, GA.

USGS. (2016). *Land Subsidence*. The USGS Water Science School. Available at: https://www.usgs.gov/special-topic/water-science-school/science/land-subsidence [Accessed on 25 June 2021].

Vershinskiy, N.V., Pshenichnyy, B.P. and Solov'tev, A.V. (1987). Artificial upwelling using the energy of surface waves. *Journal of Oceanology*, 27(3), pp. 400–402.

Volker, R.E., Mariño, M.A. and Rolston, D.E. (1985). Transition zone width in ground water on ocean atolls. *Journal of Hydraulic Engineering*, 111(4), pp. 659–676.

Volker, R.E. and Rushton, K.R. (1982). An assessment of the importance of some parameters for seawater intrusion in aquifers and a comparison of dispersive and sharp-interface modelling approaches. *Journal of Hydrology*, 56(3–4), pp. 239–250.

Voss, C.I. (1984a). *SUTRA – Saturated-Unsaturated Transport, A Finite-element Simulation Model for Saturated-unsaturated, Fluid-density-dependent Groundwater Flow with Energy Transport or Chemically-reactive Single Species Solute Transport*. U.S. Geological Survey Water-Resources Investigation Report, 84-4369. USGS, Denver, CO. Available at: https://pubs.er.usgs.gov/publication/wri844369 [Accessed on 25 June 2021].

Voss, C.I. (1984b). *AQUIFEM-SALT: A Finite-element Model for Aquifers Containing a Seawater Interface* (Vol. 84, No. 4263). US Department of the Interior, Geological Survey.

Wang, H. and Law, A.W.K. (2002). Second-order integral model for a round turbulent buoyant jet. *Journal of Fluid Mechanics*, 459, p. 397.

Wang, K.-H. (2002). Chapter 6. Linear systems approach to river water quality analysis. In Shen, et al. (eds.), *Environmental Fluid Mechanics – Theories and Application*, An ASCE Book. American Society of Civil Engineers, Reston, VA, pp. 421–457.

Worthy, J. (2003). *Large Eddy Simulation of Buoyant Plumes*, PhD. Cranfield University.

Worthy, J., Sanderson, V. and Rubini, P.A. (2001). A comparison of modified k-ε turbulence models for buoyant plumes. *Numer Heat Transfer, Part B: Fundam*, 39(2), pp. 151–166.

Wright, S.J. (1977). Mean behavior of buoyant jets in a crossflow. *Journal of the Hydraulics Division, ASCE*, 103(5), pp. 499–513.

Xu, Z., Chen, Y., Wang, Y. and Zhang, C. (2017). Near-field dilution of a turbulent jet discharged into coastal waters: Effect of regular waves. *Ocean Engineering*, 140, pp. 29–42.

Yannopoulos, P.C. (2006). An improved integral model for plane and round turbulent buoyant jets. *Journal of Fluid Mechanics*, 547, pp. 267–296.

Index

accuracy, 70–72
advection–dispersion equation
  shear flow dispersion and, 49–52
  turbulent diffusion and, 42–48
advection–molecular diffusion equation,
  30, 163
algae bloom, 13
algal bio-productivity, 217–220
alternating direction implicit (ADI)
  method, 76
artificial upwelling and mixing
  (AUMIX), 291–292, 296–297

bacterial pollution, 178
baseflow, 5
batch reactor, 105
Better Assessment Science Integrating
  Point and Nonpoint Sources
  (BASINS) modeling platform
  integration of HSPF and, 150–154
  integration of PLOAD model and,
  146–150
  unit hydrograph method, 5
biochemical oxygen demand (BOD)
  definition, 8–9
  first-stage carbonaceous wastes
  deoxygenation reaction, 168–171
  in steady-state flow, 125
biological technology, 22
buoyant plumes, 287

Canandaiqua Outlet, 208–211, 225–227
carbonaceous biochemical oxygen
  demand (CBOD)
  in Canandaiqua Outlet, 210
  deoxygenation coefficient, 215–217
  first-stage carbonaceous wastes
  deoxygenation reaction, 168–171

in modified Streeter–Phelps Models,
  186–187
in nitrification and second-stage
  deoxygenation reaction, 173
in Qingshui River, 199–203
in simulation of mass transport
  mechanisms, 53
in Streeter–Phelps model, 179, 189,
  195–196
Chakravarthy – Osher limitation, 78
Churchill formula, 18–19
Clean Water Act, 9, 10–11, 125, 134
coastal groundwater modeling
  current issues in management,
  269–272
  land subsidence, 269–270
  seawater intrusion, 270–271
  sharp interface approaches for
  seawater intrusion, 272–275
  sustainable yield, 271–272
collocation method, 85
completely stirred tank reactor (CSTR)
  batch reactor, 105
  detention ponds as, 113–119
  lake water quality systems as,
  120–124
  model, xii, 100–103
  waste-assimilative capacity analysis,
  126–128
comprehensive numerical river water
  quality models
  HEC-RAS model, 193–194
  QUAL-2K model, 90–93, 189–193
  WASP model, 193
computational fluid dynamics (CFD)
  models, 289–290
conjugate gradient method, 89–90
consistency, 72

contaminated water plumes, 231–232,
    245, 255
control volume, 27–28
convection equation schemes
    explicit central-space methods, 76–78
    implicit methods, 75–76
    oscillation control schemes, 78
    time-splitting methods, 78–79
convergence, 72
Cornell Buoyant Jet Integral Model
    (CORJET), 290
Cornell Mixing Zone Expert System
    (CORMIX), 289, 290
Courant–Friedrichs–Lewy (CFL)
    condition, 77
Crank–Nicolson method, 75–76, 79

Darcy's law, 276
deep observation wells, 285–287
deep ocean water (DOW), 290–292
deep ocean water (DOW) effluent
    plume, 291–292, 295–302
deoxygenation
    carbonaceous BOD deoxygenation
        coefficient, 215–217
    depression of river dissolved oxygen
        content by, 165–174
    first-stage carbonaceous wastes
        deoxygenation reaction, 168–171
    modified Streeter–Phelps Models,
        186–187
    nitrification and second-stage
        deoxygenation reaction, 171–174
    rate of decomposition of organic
        waste, 8–9
    simulation of mass transport
        mechanism, 53
    in steady-state flow, 125
    Streeter–Phelps model, xii, 182, 189
detention ponds, 113–119
diffusing plumes, 33–35, 39, 45
diffusion equation schemes, 79–82
diffusive mass flux, 29, 52
Dirac delta function, 31–33, 57–61,
    121–122, 213
disinfection, 8
dispersion coefficient, 285–287
dissolved oxygen (DO)
    measuring reaeration coefficient by
        DO balance, 222–225
    modified Streeter–Phelps Models,
        185–189

replenishment of DO content by
    reaeration, 174–176
river water, 8–9, 125, 157–158
Streeter–Phelps model, xii, 18,
    178–185
variation in biologically active rivers,
    176–177
water environment model of, 53
domain discretization, 83
DOW effluent plumes, 291–292, 295–302
Dupuit–Forchheimer assumptions, 272

emerging contaminants (ECs), 22–24
error minimization, 85–86
Escherichia coli (E. coli), 8, 12, 107,
    178
essentially nonoscillatory (ENO)
    schemes, 78
estuary modeling
    flushing time of well-mixed tidal
        river, 266–269
    linear system-based, 261–269
    model application, 263–266
    model formulation, 261–263
eutrophication, xii, 13, 176, 193, 220
evapotranspiration, 1–3
explicit central-space methods, 76–77

fate models, 23–24
Fick's law of molecular diffusion, 29,
    39, 45
field reconnaissance survey, 207
final survey plan, 207
final survey plan execution, 208
finite difference method (FDM)
    application in water quality models,
        14, 67
    consistency, 72
    convection equation schemes, 75–79
    convergence, 72
    diffusion equation schemes, 79–82
    order of accuracy, 70–72
    scheme construction, 69–70
    stability analysis, 72–75
    Taylor expansion, 68–69
    truncation errors, 70–72
    types, 68–82
finite element method (FEM)
    application in water quality models,
        14, 67
    background, 83
    domain discretization, 83

error minimization, 85–86
linear system of equations solution,
    87–90
matrix assembly, 86–87
shape function, 84
first-order reaction, 107–109
flood management, 4–5, 131
flow duration curve, 5
flow rate, 211
flushing time, 266–269
Flux Limiter, 78
Fourier's law of heat conduction, 29
freshwater resources, 3–4
fugacity-based multimedia fate models,
    23–24

Galerkin method, 85–86
Gauss–Seidel method, 88
geographic information system (GIS)
    application in soil transport
        modeling, 238–245
    applications, 144
    BASINS modeling platform, 146–154
    definition, 143–144
    instantaneous unit hydrograph,
        144–146
    watershed modeling, 143–154
Ghyben–Herzberg equation, 271, 274, 279
Green's function of a differentiation
    operation, 56–59
groundwater
    analytical models, 254–257
    coastal groundwater modeling,
        269–287
    dispersion coefficient, 285–287
    flow and contaminant transport in, 231
    hydrologic cycle, 1, 4
    mathematical simulation of flow and
        contaminant transport, 231–235
    model applications, 17–21
    numerical models, 253–254
    pollution, 14–17
    river water quality modeling, 5
    salinity in, 55
    SUTRA flow model, 67, 93–97
    transport simulation, 231–234,
        257–260
    water flow simulation, 234–235

Hawaii, 15–17, 135–154, 272–275,
    277, 279, 285–287
HEC-RAS model, 193–194

heuristic error analysis, 72–74
high-order accurate difference schemes,
    78–79
hydraulic parameters
    flow rate, 211
    longitudinal dispersion coefficient,
        214–215
    river velocity, 211
    time of travel, 163–164, 180, 181,
        182, 195, 196, 211–216, 221,
        226–227
hydraulic residence time distribution,
    99–100
hydrodynamic mechanisms, 53–54
hydrologic analysis, 4–6
hydrologic cycle, 1–3
Hydrologic Engineering Center-River
    Analysis System (HEC-RAS), 67
Hydrologic Simulation Program
    Fortran (HSPF) model, 5, 150–154

ideal plug flow reactor (PFR)
    ideal reactors, 103–105
    model, xii
    river water quality systems as, 125
    waste-assimilative capacity analysis,
        126–128
ideal reactors
    completely stirred tank reactor,
        100–103, 105
    concepts and mathematical
        simulation of, 99–104
    hydraulic residence time, 99–100
    ideal plug flow reactor, 103–105
    mean hydraulic residence time, 99–100
implicit methods, 75–76
indicator bacteria, 178
industrial wastewater
    minimum treatment requirements, 9
    permit system, 9–10
information technology, 22
instantaneous unit hydrograph (IUH)
    application of GIS in, 144–146
    linear systems approach to watershed
        rainfall–runoff analysis, 59, 139–143
    parameterization of, 141–143
    system parametrization, 62
    use of, 135, 139–140
integral models, 288–289
intensive river survey
    of Canandaiqua Outlet, 208–211,
        225–227

field reconnaissance survey, 207
final survey plan, 207
final survey plan execution, 208
of hydraulic parameters, 211–215
of kinetics parameters, 215–225
planning and execution, 205–211
preliminary survey plan, 206–207
procedure of, 206–208
river model formulation, 207
survey report, 208
inverse method, 61–62, 140

kinetics parameters
algal bio-productivity, 217–220
carbonaceous BOD deoxygenation
coefficient, 215–217
reaeration coefficient, 220–225

Laasonen method, 75–76
Lagrangian model, 290
lake water
completely stirred tank reactor, xii,
120–124
water quality models, 13–14
laminar flow, 41, 54
land subsidence, 269–270
Laplace transformation, 32–38, 255
lateral turbulent diffusion, 160–162
Lax method, 76–77
Lax–Wendroff method, 77, 79, 97
least-square method, 86
length-scale models, 288–289
limitation terms, 78
limiting nutrients, 13
linear reservoirs, 141
linear system-based water environment
modeling
application in soil transport
modeling, 245–253
estuary modeling, 261–269
pollution control, 121–124
river water quality modeling,
194–199
system identification in linear system
water environment modeling,
58–62
linear system of equations solution,
87–90
log-normal distribution, 62, 235, 247,
249, 252
longitudinal dispersion, 162–164

longitudinal dispersion coefficient,
214–215
Love Canal incident, 14

MA7CD10, 6, 12–13, 133–134
MacCormack method, 78–79
marine water modeling
computational fluid dynamics
models, 289–290
initial mixing of artificially upwelled
deep ocean water, 290–302
initial mixing of marine waste
discharge, 287–290
integral models, 288–289
length-scale models, 288–289
mass conservation principle, 27–30, 39
mass continuity equation, 27–29
mass transport equation, 29–30
mass transport mechanism, 52–53
mathematical formulation, 20, 21
matrix assembly, 86–87
mean hydraulic residence time, 99–100
model calibration, 21
model verification, 21
modified Streeter–Phelps Models,
185–189
modified UM model, 298–300
molecular diffusion
analytical solution of, 35–38,
45–46, 48
basic model, 31–39
Fick's law of molecular diffusion, 29,
39, 45
random walk model, 39–40
shear-flow dispersion, 52

nitrification, 171–174
nitrogenous biochemical oxygen
demand (NBOD), 53, 166, 171,
173, 179, 189, 201–202, 210, 217
nonaqueous-phase liquids (NALPs), 14
nonlinear watershed rainfall–runoff
models, 56
nonpoint source pollution
pollution problems, 13
in runoff, 12–13, 113, 131–135,
144–154
water quality models, 13
watershed modeling for, 133–135
normal distribution, 33–34, 39, 40, 42
NRFIELD, 289

numerical methods
    of coastal groundwater, 285–287
    finite difference method, 67, 68–82
    modeling of contaminant transport in
        groundwater, 253–254
    modeling of contaminant transport in
        soils, 235
    QUAL-2K, 90–93
    SUTRA flow model, 67, 93–97
    types, 67–98

O'Connor-Dobbins formula, 18–19
one-dimensional advection–diffusion
    analysis, 40–42
one-dimensional water quality model,
    39, 90, 103, 158, 160, 162
organic waste
    aerobic decomposition, 157–158,
        166, 169, 178, 184, 199
    calculating, 128
    first-order reaction, 107
    pollution control, 125
    rate of decomposition, 8–9
    settling of, 173–174
    treatment, 210
oscillation control schemes, 78

partial differential equation (PDE), 33,
    52, 57, 68, 72–73, 83, 85–86, 96,
    125, 235
permit system, 125
pesticides, 257–260
physically based water environment
    modeling
    compatibility of system-based water
        environment modeling and, 56–58
    mathematical formulation of, 54–55
    simulation of hydrodynamic
        mechanisms, 53–54
    simulation of mass transport
        mechanism, 52–53
physical parameterization, 59, 140
PLOAD model, 146–150
plumes
    buoyant plumes, 287
    contaminated water plumes, 231–232,
        245, 255
    diffusing plumes, 33–35, 39, 45
    DOW effluent plumes, 291–292,
        295–302
    pollutant plumes, 158–160, 163, 207

substance plumes, 42
    tracer plumes, 43, 214
    waste plumes, 195–196, 215, 220
point source pollution, 12, 133–134,
    153, 155, 158, 165, 186, 195
pollutant plumes, 158–160, 163, 207
pollution control
    bacterial pollution and indicating
        microorganisms, 178
    evolution of, 7–9
    linear system-based water
        environment modeling, 121–124
    point source pollution, 12, 133–134,
        153, 155, 158, 165, 186, 195
    river water, 157–158
    spatial distribution of pollutants in
        river water, 158–160
    wastewater treatment plants, 10, 12,
        22–23, 125, 133, 205, 207, 208
    water environmental modeling, 7–17
polychlorinated biphenyl (PCB)
    pollution, 14
precipitation, 1–3, 131
preliminary survey plan, 206–207

Qingshui River, 199–203
QUAL-2K model, 90–93, 189–193

rainfall
    freshwater resources, 3
    groundwater, 16
    intensity, 131–132
    linear systems approach to watershed
        rainfall–runoff analysis, 135–143
    nonlinear watershed rainfall–runoff
        models, 56, 59
    surface runoff, 4, 12
random walk model, 39–40
rate constants, 105–112
Rational Formula, 4
reaction kinetics
    effects of river reaction kinetics
        on self-purification ability, 53,
        164–178
    first-order reaction, 107–109
    laboratory determination of, 53,
        105–112
    saturation-type reaction, 110–112
    second-order reaction, 109–110
    zero-order reaction, 106–107
reaeration coefficient

measuring by tracer techniques,
221–222
measuring by the method of dissolved
oxygen balance, 222–225
reaeration rate, 8–9, 11, 18–19, 174–176
Reynolds number, 41
Richards equation, 235
river bacterial pollution, 178
river model formulation, 207
river water quality modeling
applications, 17–21
bacterial pollution and indicating
microorganisms, 178
comprehensive numerical river water
quality models, 189–194
depression of river dissolved oxygen
content by waste deoxygenation,
165–174
dissolved oxygen variation in
biologically active rivers, 176–177
effects of river hydraulic properties,
157–164
effects of river reaction kinetics,
164–178
flow duration curve, 5–6
goal of, 134
hydraulic mechanism, 11
hydrologic analysis, 4–6
ideal plug flow reactor, 125
intensive river survey, 205–227
lateral turbulent diffusion, 160–162
linear system approach, 194–199
longitudinal dispersion, 162–164
modified Streeter–Phelps Models,
185–189
QUAL-2K model, 90–93
reaeration rate, 8–9, 11
replenishment of dissolved oxygen
content by reaeration, 174–176
self-purification capacity, 5, 9, 11,
21–22, 124, 157–178
simplified river water quality models,
178–189
spatial distribution of pollutants,
158–160
Streeter–Phelps model, 178–185
surface water, 10–14
robust analytical model (RAM), 283
robust analytical groundwater flow and
salinity transport model (RAM2
[Robust Analytical Model 2]), 17,
279–285

runoff
calculating, 62
contaminated, 123, 231, 235
farmland, 208
nonlinear watershed rainfall–runoff
models, 56
nonpoint source pollution in, 12–13,
113, 131–135, 144–154
point source pollution in, 158, 210
quality, 131–133
quantity, 131–133
surface runoff, 1, 4–5
watershed rainfall–runoff analysis,
59, 135–143, 252

sag point, 182
salinity, 55
Saturated-Unsaturated Transport
(SUTRA), 67, 276–277, 279
saturation-type reaction, 110–112
seawater intrusion
definition, 270–271
sharp interface approaches for,
272–275
variable density models for, 275–279
second-order reaction, 109
shape function, 84
shear-flow dispersion, 49–52
simplified river water quality models
enhancement of hydrodynamic
mechanism simulation, 185–189
enhancement of reaction
simulation, 189
modified Streeter–Phelps Models,
185–189
Streeter–Phelps model, 178–185
sink terms, 80
soils
analytical models, 236–238
application of geographic
information system, 238–245
application of linear systems theory
in transport modeling, 245–253
application of traditional physically
based soil transport model,
246–253
flow and contaminant transport in, 231
mathematical simulation of flow and
contaminant transport, 231–235
modeling of contaminant transport,
235–245
numerical methods, 235

transport simulation, 231–234
water flow simulation, 234–235
source terms, 80
stability analysis, 72–75
stream water quality modeling, 90–93
Streeter–Phelps model, xii, 18
substance plumes, 42
subsurface contaminant transport
    modeling
    application of linear systems
        theory in soil transport modeling,
        245–253
    flow and contaminant transport, 231
    mathematical simulation of flow and
        contaminant transport, 231–235
    modeling of contaminant transport in
        groundwater, 253–257
    transport simulation, 231–234,
        257–260
    water flow simulation, 234–235
successive over-relaxation method, 89
Superfund, 14
surface water
    model applications, 17–21
    river water quality modeling, 10–14
survey report, 208
sustainable yield
    analytical models for, 279–285
    definition, 271–272
system-based water environment
    modeling
    compatibility of physically based and,
        56–58
    mathematical formulation of, 55–56
    system identification in linear, 58–62
system parametrization, 62, 140

Taylor expansion, 68–69
three-dimensional diffusion model,
    38–39
three-dimensional time-dependent
    models, 14
time of travel, 211–214
time-splitting methods, 78–79
total maximum daily load (TMDL),
    12–13, 134–135
tracer plumes, 43, 214
tracer techniques, 221–222
transport simulation, 231–234
truncation errors, 70–72, 77
turbulent diffusion, 42–48
turbulent flow, 41

two-dimensional diffusion model,
    38–39
two-dimensional time-dependent
    models, 14

UM model, 298–300
United States Environmental Protection
    Agency (USEPA), 6, 10, 125, 133,
    290, 298
United States Geological Survey,
    93, 276
unit hydrograph method, 4, 135–138
Universal Limiter for Transport
    Interpolation Modeling of the
    Advective Transport Equation
    (ULTIMATE), 78
upwind schemes, 77–78, 82
urban sewage
    minimum treatment requirements, 9
    permit system, 9–10

variable density models, 275–279
velocity
    hydrodynamic simulation, 54–55, 67
    mass transport mechanism, 29
    measuring, 211
    molecular diffusion model, 31
    one-dimensional advection–diffusion
        analysis, 41–42
    reaeration rate, 18
    river hydraulic mechanism, 11
    shear flow dispersion, 49–52
    turbulent diffusion, 43–44, 54
viscosity, 11, 41
VISJET, 290
Visual Plume model, 289–290
VISUAL PLUMES, 290
Vollenweider model, xii, 13
Volterra integral equation, 56, 58
von Neumann analysis, 74–75, 79

WASP model, 193
waste-assimilative capacity analysis,
    126–128
waste plumes, 195–196, 205, 215, 220
wastewater treatment plants (WWTPs),
    10, 12, 22–23, 125, 133, 205,
    207, 208
water flow simulation, 234–235
Water Pollution Prevention and
    Control Act, 9
Water Purification Program, 9

water quality management
  hydrologic analysis, 4–6
  lake water, 120–124
  river water quality modeling, 178,
    202–203, 205, 208
  sag point, 182
  watershed, 158
water quality modeling
  applications, 17–21
  basic procedures, 17–21
  fate and transport of emerging
    contaminates, 22–24
  mathematical formulation, 20, 21
  model calibration, 21
  model verification, 21
  with modern biological technology, 22

  with modern information technology, 22
  prospects, 21–24
  uncertainty, 17–19
watershed rainfall–runoff analysis
  linear systems approach, 135–143
  nonpoint source pollution, 133–135
  unit hydrograph method, 135–138
watersheds
  common water quality models, 5
  geographic information system and
    modeling, 143–154
  water quality analysis, 5
weighted essentially nonoscillatory
  (WENO) schemes, 78

zero-order reaction, 106–107

Printed in the United States
by Baker & Taylor Publisher Services